Agriculture Issues and Policies Series

WEEDS: MANAGEMENT, ECONOMIC IMPACTS AND BIOLOGY

AGRICULTURE ISSUES AND POLICIES

Agriculture Issues & Policies, Volume I
Alexander Berk (Editor)
2001. ISBN 1-56072-947-3

Hired Farmworkers: Profile and Labor Issues
Rea S. Berube (Editor)
2009. ISBN 978-1-60741-232-8

Effects of Liberalizing World Agricultural Trade
Henrik J. Ehrstrom (Editor)
2009 ISBN: 978-1-60741-198-7

Environmental Services and Agriculture
Karl T. Poston (Editor)
2009 ISBN: 978-1-60741-053-9

Weeds: Management, Economic Impacts and Biology
Rudolph V. Kingely (Editor)
2009 ISBN 978-1-60741-010-2

Agriculture Issues and Policies Series

WEEDS: MANAGEMENT, ECONOMIC IMPACTS AND BIOLOGY

RUDOLPH V. KINGELY
EDITOR

Nova Science Publishers, Inc.
New York

For permission to use material from this book please contact us:
Telephone 631-231-7269; Fax 631-231-8175
Web Site: http://www.novapublishers.com

LIBRARY OF CONGRESS CATALOGING-IN-PUBLICATION DATA

Kingely, Rudolph V.
Weeds : management, economic impacts and biology / Rudolph V. Kingely.
p. cm.
Includes index.
ISBN 978-1-60741-010-2 (hardcover)
1. Weeds--Control. 2. Weeds--Control--Economic aspects. I. Title.
SB611.K545 2009
632'.5--dc22
2009000157

Published by Nova Science Publishers, Inc. ✢ New York

CONTENTS

PREFACE

The term "weed" means different things to different people. In the broadest sense, it is any plant growing where it is not wanted. Weeds can be native or non-native, invasive or non invasive, and noxious or not noxious. This new book presents the latest research from around the world in this important field.

Chapter 1 - Exotic invasive plants present one of the greatest challenges to natural resource management. These weeds can alter entire communities and ecosystems, substantially degrading important ecosystem services such as forage for wild and domestic herbivores, water and soil quality, recreational values, and wildlife habitat. Traditionally, weed management in natural areas has focused on removing the target weed under the assumption that its impacts would dissipate and the system would recover following control or suppression of the invader. This approach arose presumably because most weed management tools originated in agricultural systems where removal of the weed commonly translates to increased crop yields. However, accumulating studies in natural areas indicate that simply suppressing the target weed does not ensure mitigation of its impacts. This is due primarily to the complexity of natural systems and the limitations of available management tools. To improve weed management in natural systems, we need to better understand three important factors that greatly complicate natural areas weed management: invader impacts, management side effects, and secondary invasion. Weed invasion is complex. Weeds not only impact native plants, they affect higher trophic levels, alter community interactions, and sometimes disrupt ecological processes structuring the system. Therefore, mitigating weed impacts first requires determining what the impacts are and which ones may be amenable to mitigation given current tools. Additionally, weed management tools are imprecise. As a result, management tools inevitably incur side effects on native species and system processes due to their limited selectivity and the often complex response of system components to management action. Minimizing deleterious side effects requires understanding how and why they occur. Finally, even when weed control is successful, suppression of the target weed commonly results in secondary invasion by other exotic invasive plants, thereby undermining mitigation objectives. Thus, successful natural areas weed management requires that we understand much more than just weed suppression. First, a basic understanding of the system under consideration and the impacts of the target weed on system components and processes is required to determine to what extent mitigation may be feasible. Additionally, a working knowledge of how management tools interact with the target weed, nontarget species, and system processes is necessary to maximize intended outcomes and minimize side effects of management actions. Finally, further knowledge of the ecological conditions and processes underlying successive weed invasions is needed to develop strategies for reducing the risk of secondary invasion following target invader suppression. Here, we explore these issues by examining research and management of spotted knapweed (*Centaurea maculosa*) in Western North America.

Chapter 2 - Increasing water scarcity in Asia is enforcing the adoption of water-saving rice systems such as aerobic rice, the production of direct-seeded rice in fields which are not puddled or flooded. Aerobic rice systems can greatly reduce the water requirement for rice production, but the lack of a standing water layer to suppress weed germination and growth can result in severe weed infestation. Concerns about the environmental consequences of herbicide use and the high labor costs of mechanical weeding highlight the importance of weed-competitive cultivars in the development of integrated management systems for aerobic rice. To identify crop traits related to weed-competitiveness (WC), to develop a selection protocol for WC, and to assess the role of WC in weed management, field experiments were conducted with aerobic and upland rice genotypes during 2001-2004 on the farm of the International Rice Research Institute (IRRI) in the Philippines. The research revealed that weed-suppressive genotypes are an effective component of weed management, especially when grown at an appropriate seeding rate (300 viable seeds m^{-2}). Improving weed competitiveness in aerobic rice is feasible because there exists a large genetic variability in weed-suppressive ability (WSA) among genotypes or germplasm, and because yielding ability and WSA are compatible. Fast early growth rather than plant type (erect or droopy) is the main determinant of the WSA. Indirect selection for rice yield under weed-free conditions, combined with selection for early crop vigor can simultaneously improve WSA and crop yield under weed competition.

Chapter 3 - Camelina, or gold-of-pleasure (*Camelina sativa* L. Crantz), is an oil-seed crop with a cultivation history dating back to the Neolithic era. Later, the crop was superseded by other oil seed crops such as rape or turnip rape, but during the last decade camelina has awoken new interest; recent studies show the seed oil having extraordinary good fatty-acid composition and thus being a high-value ingredient in functional food supplies. In Finland, the cultivation of camelina started in the late 1990s and the area under the crop has been growing since then; in 2006 it was cultivated on more than 5,000 ha. The weeds, however, have been found to be challenging for cultivation. As a recently introduced crop cultivated in relatively small area, there are no herbicides developed for camelina. The weeds have been controlled using herbicides such as trifluralin or metazachlor, commonly used for rape. Metazachlor is, however, known to easily injure crops belonging to the Brassicaceae family, and the use of trifluralin has been debated because of its persistence in the soil. There are also growing demand and markets for functional food supplies cultivated organically.

To develop new weed control methods for camelina, different combinations of the timing of weed control by spring-tine harrow, sowing rate and soil type were tested in a two-year field trial conducted in Helsinki, Finland. The study included two cultivars, one sown in the spring and the other in the fall. The effects of different mechanical weed control methods on the yield and above-ground biomass of the weeds were measured at-harvest, and the number of the weeds was recorded several times during the growing seasons. The effects of different mechanical weed control methods were compared with herbicide treatment with trifluralin and hand-weeded and unweeded control treatments.

The results show that the yield of camelina was more dependent on the year and the soil type than on the weed control method. In comparison with unweeded control, 0–83% of the weeds were controlled by mechanical weed control methods depending on the soil type, sowing density and the timing of the weed control practices. The efficacy of most of the mechanical weed control methods was comparable to trifluralin treatment. However, the yield in the hand-weeded control plots was higher compared with the mechanically or chemically controlled plots, indicating that the yield could be increased by weed control methods developed for camelina.

Chapter 4 - There is increasing recognition in Europe that arable plants have functions beyond simply suppressing the yield of crops. The function best quantified to date is the support they provide to higher trophic groups and to the whole food web of farmland. This food web mediates essential ecological processes such as the element cycling and pollination, and contributes aesthetic qualities through butterflies, birds, rare plants and landscape. The food web has been seriously diminished in the last half-century to the extent that it is now much less than 1% of the total summer biomass in intense arable systems. This paper summarises the evidence for the importance of arable plants to higher trophic groups and discusses the means to regenerate them with least detriment to crop yield. It shows that the field margins and the cultivated area sown with crops have their distinct contributions and should be managed integratively.

Short Communications - All living organisms, including fungi, plants, and mammals, spontaneously produce ultraweak photon emissions that originate from cellular biochemical reactions. This luminescence is commonly referred to as biophoton. The emission intensity of biophotons is extremely weak at 10^0–10^3 photons·$(s^{-1}·cm)^{-2}$, and they are considered to be associated with the oxidative metabolic responses essential for life activities, unlike the bioluminescence observed in fireflies, which employs a luciferin-luciferase system

Short Communications - Atrazine, despite being banned in most European countries, is still widely used in the United States and is registered in more than 70 countries worldwide. It is still by far the most widely used symmetrical triazine herbicide for weed control in maize in southwestern Nigeria, following which vegetable crops may be sown to 'catch' remaining moisture under rainfed multiple cropping. A randomized complete block field experiment was conducted over 2 years to evaluate the influence of residual treatments of atrazine at 1.0, 2.0, 3.0, and 4.0 kg a.i. / ha from maize cropping, including a weedy and weed-free controls on the growth and yield of amaranth at the Teaching and Research Farm of the Federal University of Technology, Akure located in the rainforest vegetation zone of Nigeria. Results indicated that atrazine at 3.0 and 4.0 kg a.i. / ha provided effective weed control and significantly ($p<0.05$) increased maize grain yield over the weedy control at a similar level with the weed-free control, but proved completely ineffective at 1.0 and 2.0 kg a.i. $ha^{-1.}$ Results further showed that atrazine at all the rates tested did not exert any adverse residual effects on the subsequent growth and yield of amaranth in both years. Consequently, it is concluded that amaranth may be sown to succeed maize in rotation without suffering a concomitant reduction in crop growth and yield under rainfed multiple cropping.

Short Communications - The anthropocentric term "weed" has been replaced by "agrestal" and "ruderal." Many agrestal and ruderal taxa are strictly confined to man-made disturbed habitats (farmland and ruderal sites, respectively) and, therefore, must have evolved in the last 12,000 years and still evolve in such places. Among these are anecophytes ("homeless plants", missing in any natural vegetation), anthropogenous taxa and inclusive cultigens. The plurality of evolutionary events resulting in anecophytes demonstrate selected examples, and implications for increasing biodiversity are discussed. Some agrestal and ruderal taxa formerly claimed to be introductions from foreign countries, exotics, are native to regions in which they have were first observed and described. In several cases, the recognition of cryptic anecophytes and cryptic species is decisive.

Short Communications - A randomized complete block field trial conducted over 2 years investigated the herbicidal efficacy of glyphosate isopropylamine salt applied singly at 1.44 kg a.e./ ha or in combination with paraquat at 0.2, 0.3, 0.4, 0.5, and 0.6 kg a.i./ha in an oil palm plantation of the Federal University of Technology, Akure, located in the rainforest vegetation zone of Nigeria. Assessment of herbicidal efficacy based on the Henderson-Tilton formula indicated that glyphosate applied alone or in combination with varying concentrations of paraquat proved highly effective in controlling most of the grasses and broadleaved weeds prevalent in the experimental plots. Herbicidal efficacy on total weed density, weed fresh weight, as well as weed dry weight was found to decrease nonsignificantly ($P<0.05$) in the order of increasing concentration levels of paraquat. It was concluded that the use of glyphosate-based paraquat mixtures at the foregoing rates could only be encouraged to improve control of susceptible glyphosate-resistant weed species in an oil palm plantation, where such weed species constitute the main weed population.

Short Communications - Concern on the safety of agrochemicals has led to more stringent international and national regulations for registration of new protection products. In an effort to maximize the chances of developing effective products with acceptable performance regarding environmental and human safety, the emphasis has been given on discovering and screening new active compounds. The input to those screens is derived from laboratory synthesized compounds but also from natural extracts with some kind of bioactivity. As far as the development of new herbicides is concerned, research on natural phytotoxic compounds has given novel chemical templates and sites of action. Several experimental protocols have been developed by research groups in order to isolate, characterize, evaluate the bioactivity and study the mode of action of natural phytotoxins. Natural phytotoxic compounds are mainly secondary metabolites of fungi or plants and are characterized by structural complexity and diversity. A large number of fungal phytotoxins have been identified and have exhibited interesting selective phytotoxicity. Unfortunately, such compounds have not fully been exploited for the development of new herbicides due to the difficulties associated with the multidisciplinary character of natural products research. Several bioactive natural compounds are structurally too complex for synthesis and low cost production. Such complex molecules do not facilitate delivery to the test-organism which is important for proper *in vitro, in vivo* and *in planta* screening. In order to overcome such obstacles research must be focused on the development of case-specific protocols and bioassays for evaluation of the herbicidal potential and elucidation of the mode of action. Other important aspects such as chemical properties, persistence and selectivity at the organism level must be examined before considering a phytotoxin as a lead compound for herbicide development. Fungal phytotoxins must be subjected to the same tests and regulation restrictions applied to synthetic phytotoxic molecules. Since the most fungal phytotoxins are produced in small quantities by phytopathogenic fungi and cause metabolic changes to affected plants, recent approaches such as genomics and metabonomics will be valuable tools in phytotoxin research.

Short Communications - One pre-emergent application of oryzalin 40% SC herbicide and one hoeing was found to control maximum pests of cabbage crop. By following proper date of sowing of cabbage crop and controlling the weeds, no pest including insect and diseases was observed to attack the crop. During the period of controlling weeds of cabbage crop neither insect pest nor any fungal attack was observed during the period of this study. The insects and diseases were thus below economic threshold value.

Short Communications - Since man first began to cultivate crops, undesirable plants called weeds have been a problem. Weeds have long been recognized as major pests causing significant losses in crop productivity (Bhowmick and Ghosh, 2002a; Bhowmick and Biswas, 2007), because, it is unfortunate that the damage on crops by weeds is invisible as compared to other pests. Farmers appreciate weeds as a problem only when it goes out of hand, or when they seriously affect agricultural operations such as harvesting (e.g. spiny weeds). In addition to direct effect on crop yield, weeds result in considerable reduction in the efficiency of inputs used. The precious and costly inputs such as fertilizers and irrigation water, which are otherwise meant for realizing the potential yield, are depleted by the weeds (Yaduraju and Mishra, 2005; Bhowmick and Biswas, 2007).

In: Weeds: Management, Economic Impacts and Biology
Editor: Rudolph V. Kingely
ISBN: 978-1-60741-010-2

Chapter 1

MANAGING INVASIVE PLANTS IN NATURAL AREAS: MOVING BEYOND WEED CONTROL

Dean Pearson and Yvette Ortega
Rocky Mountain Station, USDA Forest Service,
800 East Beckwith Ave., Missoula, MT 59801, USA

ABSTRACT

Exotic invasive plants present one of the greatest challenges to natural resource management. These weeds can alter entire communities and ecosystems, substantially degrading important ecosystem services such as forage for wild and domestic herbivores, water and soil quality, recreational values, and wildlife habitat. Traditionally, weed management in natural areas has focused on removing the target weed under the assumption that its impacts would dissipate and the system would recover following control or suppression of the invader. This approach arose presumably because most weed management tools originated in agricultural systems where removal of the weed commonly translates to increased crop yields. However, accumulating studies in natural areas indicate that simply suppressing the target weed does not ensure mitigation of its impacts. This is due primarily to the complexity of natural systems and the limitations of available management tools. To improve weed management in natural systems, we need to better understand three important factors that greatly complicate natural areas weed management: invader impacts, management side effects, and secondary invasion. Weed invasion is complex. Weeds not only impact native plants, they affect higher trophic levels, alter community interactions, and sometimes disrupt ecological processes structuring the system. Therefore, mitigating weed impacts first requires determining what the impacts are and which ones may be amenable to mitigation given current tools. Additionally, weed management tools are imprecise. As a result, management tools inevitably incur side effects on native species and system processes due to their limited selectivity and the often complex response of system components to management action. Minimizing deleterious side effects requires understanding how and why they occur. Finally, even when weed control is successful, suppression of the target weed commonly results in secondary invasion by other exotic invasive plants, thereby undermining mitigation objectives. Thus, successful natural areas weed management requires that we understand much more than just weed suppression. First, a basic understanding of the system under consideration and the impacts of the target weed on system components and processes is required to determine to what extent mitigation may be feasible. Additionally, a working knowledge of how management tools interact with the target weed, nontarget species, and system processes is necessary to maximize intended outcomes and minimize side effects of management actions. Finally, further knowledge of the ecological conditions and processes underlying successive weed invasions is needed to develop strategies for reducing the risk of secondary invasion following target invader suppression. Here, we explore these issues by examining research and management of spotted knapweed (*Centaurea maculosa*) in Western North America.

INTRODUCTION

Exotic invasive plants (also referred to here as weeds) present a formidable threat to the economic productivity and ecological integrity of natural systems (Mack et al. 2000; Pimentel et al. 2005). As a result, weed management is now a primary objective for many federal, state, and private land stewards (e.g., Randall 1996; Hulme 2006). However, achieving successful weed management in natural areas has proven elusive and unpredictable (D'Antonio et al. 2004; Smith et al. 2006). One potential reason for this is an overly narrow focus on controlling the target weed. Historically, management strategies have focused on removing or suppressing the invader under the assumption that its negative impacts would dissipate in its absence or reduced vigor (Hobbs and Humphrey 1995). Presumably, this approach arose because most weed management practices originated in agricultural systems where controlling the target weed often results in a direct increase in the productivity of the desired crop plant (Smith et al. 2006). However, in natural areas, controlling the target weed commonly fails to result in system recovery (e.g., Story et al. 2006; Ortega and Pearson in review). Addressing this problem requires explicitly recognizing how differences between natural areas and crop systems may influence management outcomes so we can better adapt strategies for natural areas management.

The fundamental distinction between natural areas and crop systems is ecological complexity. Natural areas contain a great variety of species whose composition, productivity, and relative abundance are determined by an array of community interactions and system-level processes. As a result, weed management in natural areas involves management of entire communities, not just individual weeds. That is, natural areas weed management is in essence *directed community assembly* (Holt and Hochberg 2001; Pearson and Callaway 2005). As such, it requires that we understand community- and system-level processes and how invasion and management actions influence these processes in order to determine specifically which weed impacts may be mitigated and how.

Prior weed management assessments have evaluated standard practices, explored theoretical frameworks, and offered guidelines for integrated approaches that have advanced both our thinking and applications in this field (Hobbs and Humphries 1995; Randall 1996; Sheley et al. 1996, 1998, 2006; DiTomaso 2000; D'Antonio et al. 2004; Smith et al. 2006; Hulme 2006). Several such assessments have indicated the importance of setting forth clear weed management objectives that take into account larger land management objective before initiating weed management actions (e.g., Randall 1996; Sheley et al. 1998; Rew et al. 2007). However, before we can effectively identify the feasibility of weed management actions, we must determine the nature and extent of weed impacts on the system and understand how management actions can influence overall outcomes through their interactions with the weed and other system components and processes. Here, we focus on these issues in an effort to move natural areas weed management beyond simple weed control toward a more holistic discipline.

Although weed management in natural areas has become quite prevalent in recent years, a formal definition of natural areas for weed management purposes is lacking. This situation has hindered natural areas weed management by not clearly distinguishing issues in natural areas from those associated with agricultural systems. For our purposes, we define natural areas as non-crop systems with residual native vegetation where management objectives include retaining or augmenting native vegetation to generate ecosystem services. This definition encompasses a range of conditions from virtually pristine areas with few weeds to heavily invaded or otherwise degraded areas that still contain some semblance of the original natural vegetation. Thus, urban environments and cropland systems are excluded, but most timber production lands, recreational lands, and grazing lands are included where native vegetation is intended to generate some or all of the primary productivity or other ecosystem services such as soil stability, water conservation, wildlife populations, natural aesthetics, etc. With this definition in hand, we can begin to discuss the complexities of natural areas weed management as a distinct discipline.

In many regards, natural areas management is analogous to the practice of human medicine. In human medicine we use a variety of physical and chemical treatments in an effort to alleviate ailments within complex biological organisms. Due to the sheer complexity of the human body, and in many cases insufficient precision of the drugs and treatments used, virtually every medical treatment has a cadre of side effects, which may range in severity from minor to quite serious (e.g., Evans and McLeod 2003; Niggemann and Gruber 2003). Thus, successfully managing human health requires the weighing of potential treatment side effects against the intended treatment outcome for a given illness. The same is true in natural areas management. Within natural systems, all management actions have side effects due to the complexity of natural systems and the limited specificity of the tools employed. Management side effects can vary from innocuous to rather severe, and, as with human medicine, the severity of the illness determines the acceptable risk of side effects. Thus, effective resource management requires balancing management side effects against the intended outcome for a given management objective. Accomplishing this requires a thorough understanding of the potential range of side effects associated with different resource management tools.

As the history of medicine shows, advancing an applied science involves evolution by trial and error. In ancient Peru, France, and Egypt, a primitive form of brain surgery was developed called trepanation (Arnott et al. 2003). This practice involved physically opening the skull with primitive metal tools and was believed to be used for a variety of purposes ranging from treating head injuries from battle to curing epilepsy and relieving headaches. Modern forensic analysis indicates many patients survived and often received multiple treatments, suggesting the practice was successful. Today trepanation might seem to be an extreme treatment for headaches, with potentially excessive side effects, given contemporary medicine. Modern medicine has come a long way since the early practices of trepanation through the development of better understandings of the human body, the diseases that ail it, and refinement of the tools and techniques we use as treatments. In contrast, natural areas weed management is in many ways still in its infancy. To date, we have developed relatively few effective tools, they tend to exhibit limited selectivity, our understandings of particular "diseases" such as biological invasions are still quite primitive, and although we are beginning to gradually improve knowledge of natural systems, we are not particularly efficient at integrating this new knowledge into applied management approaches (D'Antonio et al. 2004; Smith et al. 2006). Thus, in many regards contemporary weed management may be likened to ancient trepanation – we commonly deploy tools that have rather severe side effects largely because we lack more precise tools and have insufficient knowledge of the complex systems we work in and the diseases or threats we wish to abate. This is not intended as a casual criticism, but rather as an objective assessment of the current state of this relatively young science.

At this stage in our evolution, we need to further refine the science in an effort to develop more surgical precision and improve the field as a whole. We acknowledge and accept that the state of any science will progress through trial and error. However, much as the Hippocratic Oath guides the ethic of human medicine, we feel that the guiding principal for invasives management should be "*to do no harm.*" That is to say, management actions should improve the system as a whole, as measured by the overall system response to management and not simply by the degree to which the target weed is suppressed. As we illustrate below, using tools without carefully considering the collective outcome can actually make conditions worse (e.g., Pearson and Callaway 2003; Lenhoff et al. in press; Ortega and Pearson in review). Expending limited funds and resources to make matters worse is not an acceptable outcome. To improve natural areas management we need to better understand the ecology of the systems we are working in, the nature of the threats we wish to mitigate, and the efficacy of the tools we employ. In this context, we examine three major factors that must be understood to improve natural areas weed management: invader impacts, management side effects, and secondary invasion.

Effective weed management in natural areas begins with understanding the nature and extent of weed impacts in the invaded system. When an exotic invades a natural area, it develops numerous interactions with system components. These interactions may be positive, negative, trophic, nontrophic, biotic, abiotic, direct and indirect. The overall result is that strong invaders (*sensu* Ortega and Pearson 2005) do not just reduce the abundances of individual native organisms, they may alter community interactions, abiotic resources and/or physical conditions in ways that restructure the community and sometimes alter fundamental ecological processes such as nutrient cycling, hydrologic conditions, and disturbance regimes (e.g., Vitousek 1990). Notably, the specific impacts that an invader has may differ across communities (Rew et al. 2007; Maxwell et al. 2008), requiring that we understand the mechanisms underlying observed changes. Thus, effectively mitigating invader impacts requires understanding how the invader alters not only abundances of individual species but also community interactions and the processes structuring the system.

A major obstacle to successful weed management in natural areas is the fact that management tools inevitably have side effects. This is due to the limited selectivity of management tools and the complexity of the organisms and processes they interact with. Because management tools can themselves be biotic or abiotic, pulsed or persistent, widespread or localized, and broad or narrow in their intended selectivity, they can exhibit a wide array of side effects in natural systems. These side effects can further vary as a function of local site conditions (Shea et al. 2005; Ortega and Pearson in review). Side effects of management actions include reducing vigor or abundance of native or desirable species, inhibiting overall productivity or diversity, shifting community structure and function, altering physical conditions, and more (Louda et al. 1997; D'Antonio et al. 2004; Smith et al. 2006; Hulme 2006; Ortega and Pearson in review). Therefore, effective weed management requires weighing the success of control measures (e.g., impacts on target weeds and recovery of native species) against the side effects of management actions. This necessitates a thorough understanding of how management tools interact with nontarget system components as well as the target weed (e.g., Pearson and Fletcher 2008).

Weed management in natural areas is further complicated by secondary invasion. Secondary invasion occurs when successful control or suppression of the target weed results in proliferation of another exotic invader. In many cases, the secondary invader may cause more problems and present a greater management challenge than the original target weed (Hulme 2006; Ortega and Pearson in review). Although we note that secondary invasion could be considered a management side effect, we treat it as a separate condition because it involves additional species extrinsic to the original native system. Thus, we treat management side effects as the unintended effects of the management action on native system components or processes, whereas we treat secondary invasion as a compensatory response of the exotic subset of the community to suppression of the target weed. We feel this distinction is helpful because secondary invasion is not so much a side effect of a specific management tool on nontarget species as it is a failure or incapacity to address the weed problem itself, i.e., the system-level processes such as propagule pressure or disturbance underlying and promoting invasions (e.g., Hobbs and Huenneke 1992; Davis et al. 2000; Seabloom et al. 2003; McDougal and Turkington 2005; Smith et al. 2006; Thomas and Reid 2007). Readers not fully comfortable with this distinction may choose to think of secondary invasion as a special case of management side effects, albeit one that is sufficiently important to warrant special attention.

We explore these challenges in natural areas weed management by examining spotted knapweed (*Centaurea maculosa*) invasion and management in the western United States and supplementing this case study with complementary research on other invaders. *Centaurea maculosa* has been identified as the single most studied terrestrial invasive plant (Pysek et al. 2008) and is prominent among a group of knapweeds identified as among the most serious plant invaders in the western United States (Sheley et al. 1998; LeJeune and Seastedt 2001). *C. maculosa* is a perennial, tap-rooted forb in the Asteraceae family that was first discovered in North America in the late 1800s (Strange et al. 1979) and is thought to have arrived from Europe as a contaminant of alfalfa seed (Groh 1944). It is an aggressive invader that rapidly spread across the United States and has become a particular problem in semi arid habitats of the western United States (Sheley et al. 1998).

Several studies have examined a broad range of *C. maculosa*'s impacts and interactions in natural systems (Tyser and Key 1988; Ridenour and Callaway 2001; Ortega and Pearson 2005; Ortega et al. 2006; Pearson 2009), while others have explored potential mechanisms to explain its successful invasion (Ridenour and Callaway 2001; Callaway and Ridenour 2004). Its infamy has made *C. maculosa* a primary target of weed management using biocontrol, herbicides, and grazing (Harris and Myers 1984; Rice et al. 1997a; Olson et al. 1997; Sheley et al. 1998). Importantly, *C. maculosa* management outcomes have been some of the most thoroughly studied for weed management of natural areas (Thompson 1996; Rice et al. 1997a; Rice and Toney 1998; Tyser et al. 1998; Pearson et al. 2000; Ortega et al. 2004; Olson et al. 1997; Story et al. 1995, 2006, 2008; Pearson and Fletcher 2008; Pearson and Callaway 2003,

2006, 2008; Ortega and Pearson in review). Thus, *C. maculosa* offers a comprehensive case study for examining invasive species management. Here, we examine *C. maculosa*'s impacts on and interactions with native system components, the efficacy and side effects of major management tools, and the extent and severity of secondary invasion to better understand the challenges of natural areas weed management.

C. MACULOSA IMPACTS AND INTERACTIONS IN THE INVADED SYSTEM

If simply controlling weed populations in natural areas resulted in full restoration of all system components, there would be no need to assess weed impacts. We could simply focus on suppressing weed populations, because this would fix the problem. However, as highlighted in later sections, weed management actions rarely mitigate all weed impacts, and they virtually always have side effects on nontarget species. Therefore, understanding weed impacts and interactions within the invaded system is prerequisite to assessing which impacts may be amenable to which mitigation efforts. Yet, despite numerous studies documenting the profound ecosystem-level effects strong invaders can have (e.g., Vitousek 1990; Levine et al. 2003), there are surprisingly few cases where a weed's impacts on and interactions with native species and community processes have been examined in detail. The case of *C. maculosa* invasion in the western United States offers one of the best such examples. Extensive research indicates that *C. maculosa* has a range of positive and negative direct and indirect interactions with native species that have altered community interactions at multiple trophic levels and have important implications for its management. We briefly discuss these results here.

At the lowest trophic levels, evidence suggests that *C. maculosa* substantially suppresses a range of native plant species (Tyser and Key 1988; Ridenour and Callaway 2001; Ortega and Pearson 2005, in review). However, not all species are equally at risk. A detailed analysis of the relationship between *C. maculosa* and native plant functional groups suggested that *C. maculosa* had the strongest negative effects on native perennial forbs, moderate negative effects on perennial grasses, and no detectable effect on annual forbs (Ortega and Pearson 2005). These results suggest that *C. maculosa* invasion initiates a trajectory within these systems that shifts the plant community by suppressing perennial forbs and grasses that typically dominant native biomass. Since forbs are the primary source of diversity, commonly making up ≥80% of species richness in these systems (Mueggler and Stewart 1980; Pokorny et al. 2005), *C. maculosa* invasion also takes a toll on native plant diversity (Ortega and Pearson 2005). This impact trajectory has important implications for management. Ideally, management actions would be directed at reversing the trajectory, but limitations in the selectivity of management tools may result in management efforts that actually exacerbate certain invader impacts. For example, broadleaf herbicides commonly suppress native forbs (Rice et al. 1997a; Crone et al. 2009; Ortega and Pearson in review), potentially exaggerating trajectories initiated by *C. maculosa* invasion even if the invader is successfully removed (see *Efficacy and side effects of management actions* below).

Most studies of exotic plant impacts do not extend beyond lateral plant-plant interactions to examine how plant invasions affect higher trophic levels, particularly at the level of predators. Most efforts to assess higher trophic level effects have been limited to extrapolating forage production to estimate impacts on domestic or wild ungulates (e.g., Trammell and Butler 1995). *C. maculosa* is believed to reduce forage for domestic and wild ungulates through its direct reductions in native plant biomass (Thompson 1996; Rice et al. 1997b). However, native ungulates do use *C. maculosa* as forage to some extent (e.g., Wright and Kelsey 1997), and no studies to our knowledge have examined the effects of *C. maculosa* or other invasive plants on population-level responses of large herbivores. This is undoubtedly due to the difficulty of conducting studies at the necessary scale to effectively accomplish this. Nonetheless, it is reasonable to assume that there is some threshold of *C. maculosa* abundance, and therefore reductions in native plant biomass, beyond which large wild herbivores such as deer (*Odocoileus* spp.), elk (*Cervus elaphus*), and sheep (*Ovis canadensis*) might experience population-level impacts, which may also affect higher trophic levels.

One study provides evidence that weed-induced reductions in native plant biomass and diversity impact not only herbivores, but also higher trophic levels by eroding food chains (Ortega et al. 2006). Pitfall sampling in *C. maculosa* invaded and comparable uninvaded sites suggests that *C. maculosa* invasion reduced biomass and richness of invertebrate groups such as grasshoppers (Orthoptera), which are important food resources for many small predators such as birds, small mammals, and spiders. In this scenario, *C. maculosa* invasion and resulting declines in invertebrates were associated with negative impacts on an insectivorous songbird, the chipping sparrow (*Spizella passerina*), as indicated by delayed nest initiation, lower site fidelity of breeding adults, reduced likelihood of double brooding, and, overall, projected lower fecundity for sparrows on *C. maculosa*-invaded sites (Ortega et al. 2006). Because chipping sparrow nests are located in trees, which are unaffected by *C. maculosa* invasion, and nest predation did not vary between invaded and uninvaded habitats, *C. maculosa* impacts on chipping sparrow fecundity appear to arise through reductions in food for parents and possibly also offspring. In addition to showing that *C. maculosa* can impact invertebrates and other wildlife, these results illustrate how invasion by exotic plants like *C. maculosa* may erode native food chains by reducing herbivore populations and impacting the predators they support. This suggests that removing the weed without restoring the native plant biomass and diversity, may fail to restore certain herbivores and predators due to residual impacts on food chains.

Exotic plant invasions may affect higher trophic levels and trophic interactions through less obvious means as well. Invasion of western North American grasslands by *C. maculosa* and other European forbs is shifting the vegetation architecture of these grasslands, with substantial implications for predator populations and predator-prey interactions (Pearson 2009). European forbs like *C. maculosa* that are invading these grasslands are generally larger, more rigid, and more structurally complex compared to the native grassland vegetation. This has lead to a dramatic increase in native web-building spiders by providing a superabundance of new substrates favorable for web construction. It has also affected the hunting behavior of these predatory spiders by allowing them to become more efficient. Webs of *Dictyna* spp. spiders constructed on *C. maculosa* plants were three to four times larger and captured two times more total prey and four times more large prey compared with webs on native substrates. The overall result was that *C. maculosa* increased predation rates on spider prey by ≥89 fold in invaded grassland due to the cumulative effects of *C. maculosa* architecture on spider density and hunting behavior. This research shows that exotic plant invasions can influence higher trophic levels not only through food chains as described above for chipping sparrows, but also through changes in vegetation architecture that directly affect predator densities by altering their habitat. Additionally, because plant architecture provides the arena for predator-prey interactions, such changes can have complex effects on predator-prey dynamics. It also shows that invasive plants can have substantial positive effects on populations of some native species, as noted elsewhere (Rodriguez 2006; White et al. 2006). However, due to indirect interactions, even very strong positive direct effects of invasion on some species may lead to very strong negative effects on other native species such as spider prey (Pearson 2009).

In addition to biotic effects, *C. maculosa* may affect abiotic components of native systems as well. For example, *C. maculosa* appears to increase soil erosion (Lacey et al. 1989), a condition that could result in long-term residual damage to a site even if *C. maculosa* is completely removed. *C. maculosa* also produces chemicals such as the root exudate, ±catechin (Bais et al. 2003), that can have phosphorus chelating effects within soils (Callaway and Ridenour 2004). Although catechin is not thought to be particularly persistent in the absence of *C. maculosa*, possible effects on nutrient availability could persist for unknown periods after the weed's removal. In general, exotic species that alter soil chemistry, soil structure, or hydrologic conditions may have residual community- and ecosystem-level effects that persist following control.

Collectively, these studies show that *C. maculosa* affects not only plant communities, but also soils, consumers, predators, and important community processes through positive, negative, direct, indirect, biotic and abiotic interactions. Because most *C. maculosa* impacts on the native system arise as a result of its effects on vegetation, controlling *C. maculosa* populations is only likely to restore preinvasion conditions to the extent that native vegetation recovers (one exception is *C. maculosa*'s direct effects on spiders and indirect effects on spider prey, which may be largely reversed by simply removing the weed). Recovery of the plant community may in turn depend on the ability to mitigate abiotic impacts of invasion, restore seed sources, and address similar factors that may interfere with native plant recovery (e.g., Seabloom et al. 2003; McDougal and Turkington 2005). Although full restoration may represent the Holy Grail of natural areas weed management, due to the extent and complexity of interactions that exotic plants may establish in natural systems and their own pervasiveness, full restoration following heavy invasion is rarely feasible. On more intensively managed natural areas such as grazing lands, it may not even be desirable, and it is often not the primary objective. Nonetheless, the more weed impacts can be mitigated, the more ecosystem services are likely to be restored to preinvasion conditions. This should be the gold standard for managing pubic lands.

Efficacy and Side Effects of Management Actions – A Balancing Act

The intended effect of a management tool on the target weed is only one aspect of its influence on the treated system. As with human medicine all treatments have side effects, and successful management requires finding an appropriate balance between negative side effects of treatment and the intended treatment response. To accomplish this, we need to understand what kinds of side effects can arise from different management tools.

Limited selectivity of weed management tools is perhaps the greatest cause of side effects and therefore one of the greatest limitations to improving management in natural areas. However, side effects can also arise from highly selective control measures due to indirect interactions involving nontarget species (Pearson and Callaway 2003). Additionally, side effects can arise through physical factors such as when management tools such as grazing or fire create disturbances. Duration or persistence of a treatment can determine the strength and therefore the biological significance of its side effects. For example, the lack of selectivity of an herbicide may not result in biologically significant side effects on nontarget species if the effect is sufficiently short-lived. However, the same side effect may become biologically significant if it persists over time (Crone et al. 2009). Here we evaluate management tools and how they interact with natural systems to better understand how deleterious side effects of management actions can be avoided or minimized. We do so by focusing on *C. maculosa* management using some of the more prominent tools employed for weeds in natural areas: herbicide, classical biological control, and grazing. Because secondary invasion can be an important consequence of any weed management measure, we address issues of secondary invasion separately in the next section.

Herbicide

Herbicide applications are increasingly used in natural areas weed management, with an emphasis on broadleaf herbicides for controlling exotic forbs (DiTomaso 2000). Broadleaf herbicides are heavily used for *C. maculosa* control and extensive research has been conducted to evaluate the efficacy of this method and identify ways to minimize its side effects on native plants (Rice et al. 1997a; Rice and Toney 1998; Tyser et al. 1998). These studies show that *C. maculosa* is fairly sensitive to several broadleaf herbicides. As a result, herbicides can be used at low dosages in an effort to minimize side effects on nontarget plants while still achieving effective *C. maculosa* control (≥80% reductions of the target species for several years; Rice et al. 1997a; Sheley and Jacobs 1997; Rice and Toney 1998). Additionally, because *C. maculosa* is actively growing later than many native forbs, selectivity can be enhanced by applying herbicides after other forbs have senesced (Rice et al. 1997a). Yet, despite these refinements, broadleaf herbicide treatments for *C. maculosa* can still have serious side effects (Ortega and Pearson in review).

The use of broadleaf herbicides for *C. maculosa* control has been shown to shift the composition of native communities away from forbs toward grasses (Ortega and Pearson in review). This is of particular significance in intermountain grasslands where *C. maculosa* is most problematic because forbs generally comprise ≥80% of the species richness in these systems and can dominate biomass (Mueggler and Stewart 1980; Ortega and Pearson 2005; Pokorney et al. 2005). It is also of concern because, as discussed above, *C. maculosa* invasion itself shifts community composition away from dominance by perennial forbs (Ortega and Pearson 2005). Thus, if not carefully used in this system, broadleaf herbicides may exacerbate certain invader impacts even if they control the target weed.

Herbicide-driven declines in forbs can potentially be substantial due to demographic effects. In addition to suppressing cover of adult native forbs, broadleaf herbicides can kill young plants and suppress reproduction for several years following treatments (Crone et al. 2009; Ortega and Pearson in review). Such effects can create population bottlenecks for native forbs that can result in demographic declines over time. Long-term demographic analysis suggests that the degree to which such impacts may threaten persistence of a species in herbicide treatment areas will depend on the persistence of the herbicide and the reapplication rates required to maintain control of the target weed relative to the life history characteristics of the plant (Crone et al. 2009). These efficacy studies offer suggestions for further refining broadleaf herbicide applications to reduce side effects in natural areas. Emphasizing spot spraying of target weeds over broadcast spraying can reduce the spatial extent of side effects and encourage recovery of native species by providing vigorous seed sources in uninvaded patches that are protected from herbicides. Additionally, target weed recovery should be carefully monitored to ensure that re-treatment to maintain control of the target weed is not premature. This will reduce the risk of excessively reapplying herbicides at rates that can further depress native forb populations (Crone et al. 2009).

Most of the lessons learned from studies of *C. maculosa* herbicide management will also apply to management of other exotic forbs using broadleaf herbicides. However, many other exotic forbs are less sensitive to herbicides and require stronger dosages (Rice and Toney 1998; Tu et al. 2001), so the types of side effects described here will tend to be stronger and more difficult to mitigate. Although side effects such as loss of forbs may be acceptable for relatively restricted areas under specific management objectives such as intensive forage production for domestic and wild herbivores, large scale reductions of forb diversity in grassland systems should be a major concern for natural areas management. Forbs commonly represent the bulk of the species and functional diversity in natural grasslands (Seabloom et al. 2003; Pokorney et al. 2004). They support communities of pollinators, herbivores, and higher trophic levels (e.g., Potts et al. 2003; Ortega et al. 2006). They also may be an important factor in maintaining resistance of natural areas to future invasion (Dukes 2002; Zavaleta and Hulvey 2004; Maron and Marler 2007).

Grazing

Grazing for the purpose of controlling range weeds has gained interest and application over the past 10-15 years (Olson et al. 1997; Sheley et al. 1998; DiTomaso 2000). Grazing is generally less effective at controlling weeds than herbicides because it does not usually kill plants outright but rather reduces biomass, flowering, and seed production (Olson et al. 1997; Sheley et al. 1998). Like herbicide applications, grazing tends to target exotic forbs by using domestic herbivores such as sheep (*Ovis aries*) or goats (*Capra aegagrus hircus*) that either have a natural affinity for the target weed or can be trained to feed on the target weed (e.g., Walker et al. 1994; DiTomaso 2000). Within intermountain grasslands, the fact that many exotic forbs are green after most natives have begun to senesce is exploited to increase selectivity by timing grazing so animals will focus on the actively growing and hence more palatable target weed (Olson et al. 1997). To be most effective, grazing must be short in duration to minimize disturbance from trampling and reduce the tendency of animals to forage on nontarget species. The numbers of animals grazed must also be minimized for similar reasons. Thus, grazing to reduce weeds requires careful oversight to ensure it is effective. Depending on the target weed and the ability to control the timing and specificity of the foraging, grazing may be more or less selective than broadleaf herbicides. Use of forb eaters like sheep can result in reductions in forbs and increases in grasses similar to herbicides (R. Kott, Montana State University, Bozeman, MT, pers. comm.). In this regard, grazing can push a community away from grasses and exacerbate the trends begun by the weed itself. Excessive grazing is well known to facilitate weed invasion (e.g., Loeser et al. 2007), so if grazing for weed management is not *very* carefully conducted, it will almost certainly exacerbate the weed problem.

Biological Control

Classical biological control (referred to hereafter as biocontrol) involves the introduction of exotic organisms, usually invertebrates from the native range of the invader, to suppress pest populations through top-down control (Thomas and Reid 2007; Müller-Shärer and Schaffner 2008). When successful, biocontrol can be an extremely powerful and economical tool over large areas (Gurr and Wratten 2000). As a result, biocontrol is perhaps the most widely used weed management technique in natural areas (DiTomaso 2000; Coombs et al. 2004). The history of biocontrol illustrates not only the importance of selectivity of weed management tools, but also the limitations. Historically, biocontrol was practiced using generalist species including vertebrate predators and polyphagous insects that resulted in extensive side effects when they attacked nontarget species (Howarth 1991; Simberloff and Stiling 1996; Louda et al. 1997; Henneman and Memmott 2001). To address this problem, weed biocontrol has become progressively more stringent in selecting for highly specialized agents that attack only the target weed (McEvoy 1996). Today, weed biocontrol agents are highly host-specific (Pemberton 2000), and biocontrol of weeds is arguably the single most selective weed management tool currently in use aside from hand pulling. However, recent examination of *C. maculosa* biocontrol has established that even highly host-specific biocontrol agents can have rather significant nontarget effects, indicating a need to better understand these side effects to guard against them (Pearson and Callaway 2003, 2005).

Until recently, host-specificity was deemed the ultimate assurance that exotic organisms introduced for biological control would not harm nontarget species. However, studies now show that even if a biocontrol agent is completely host-specific, it can still exhibit nontarget effects through food-web interactions simply by being consumed by other organisms (Pearson et al. 2000; Pearson and Callaway 2003). Although, it is well documented that exotic invertebrates introduced for biological control are commonly consumed by a variety of native and nonnative species, such interactions had previously been studied only as potential sources of biocontrol interference (Goeden and Louda 1976). Community-level examination of *C. maculosa* biocontrol shows how such consumer interactions can bring about indirect nontarget effects.

The gallflies, *Urophora affinis* and *U. quadrifaciata*, introduced for *C. maculosa* biological control, are eaten by numerous native generalist species (Story et al. 1995; Pearson et al. 2000). Several studies suggest that gallfly consumption by native deer mice (*Peromyscus maniculatus*) results in food subsidies that elevate deer mouse populations by two to three fold (Ortega et al. 2004; Pearson and Callaway 2006; Pearson and Fletcher

2008). Moreover, gallfly–subsidies to deer mice have been correlated with higher incidence of Sin Nombre hantavirus in deer mouse populations (Pearson and Callaway 2006). Because Sin Nombre hantavirus is the etiological agent for the deadly hantavirus pulmonary syndrome in humans (Childs et al. 1994), elevating hantavirus over large regions of the landscape has potentially significant implications for the ecology and epidemiology of this disease. Biocontrol food subsidies to deer mice also appear to negatively impact some of the native plants they were introduced to benefit. As efficient seed predators, deer mice can reduce recruitment of juveniles and, ultimately, abundance of adults in native plant populations through seed predation (Maron and Kauffman 2006; Bricker et al. in press). A recent study in western Montana showed that deer mice reduced recruitment of the dominant native grass, *Pseudoroegneria spicata*, and forb, *Balsamorhiza sagittata*, through seed predation and this effect was significantly greater for *B. sagittata* where biocontrol food subsidies elevated mouse populations (Pearson and Fletcher 2008). These results suggest that gallflies that were introduced to reduce *C. maculosa*'s impacts on native plants may actually be increasing *C. maculosa*'s negative impacts on some native plants through food subsidies via second-order apparent competition. Ongoing work suggests that gallfly food subsidies to black-capped chickadees (*Poecile atricapillus*) may influence foraging behavior and competitive interactions among chickadee species and possibly other songbirds (L. Greenwood, unpublished data). In contrast, other work suggests that spiders that live on *C. maculosa* and consume adult *Urophora* are far more affected by structural attributes of *C. maculosa* than food resources provided by *Urophora* (Pearson 2009).

The study of *C. maculosa* biocontrol illustrates that host-specific agents can have significant impacts on nontarget species through food-web interactions. Understanding how such interactions take place is crucial to mitigating these side effects in the future. The strength of the indirect effects originating from *Urophora* biocontrol agents may be attributed to several factors: *Urophora* are highly palatable, they are readily available over extended periods of time, they are super-abundant, they are exploited by strongly interacting generalist consumers, and they have not controlled the target weed. If one or more of these criteria are not met or greatly diminished, indirect effects of a biocontrol agent arising through food-web interactions will likely be attenuated or weakened. Even if a biocontrol agent meets all of the criteria described above, as *Urophora* do, any indirect food-web effects may be rendered ephemeral and biologically unimportant if the control agent ultimately suppresses the target species (Pearson and Callaway 2003). This is because the host-specificity of the biocontrol agent will ensure that it too declines by eating itself out of house and home as it reduces its host populations.

Recent reports of local reductions in *C. maculosa* populations have been attributed to successful biocontrol of this weed (Story et al. 2006, 2008) and interpreted as an example of biocontrol success ultimately nullifying indirect nontarget effects arising from gallfly subsidies to deer mice (Van Driesche et al. 2008). Hopefully, these assertions will ultimately be born out, but they are currently premature. Long-term monitoring shows that the key agent postulated as controlling *C. maculosa*, the root weevil (*Cyphocleonus achates*), is absent at many locations where *C. maculosa* has recently declined (Sturdevant et al. 2006). Most other *C. maculosa* biocontrol agents are also absent from numerous sites where *C. maculosa* has declined (Sturdevant et al. 2006). Thus, biocontrol alone cannot explain these observations.

Reductions in spring precipitation offer the most parsimonious explanation for widespread *C. maculosa* declines. Spring precipitation is highly correlated with *C. maculosa* declines and is the only factor consistently present at all locations where declines have been reported (Pearson and Fletcher 2008; D. E. Pearson unpublished data). Notably, Sturdevant et al. (2006) did report trends toward more severe declines in *C. maculosa* vigor at sites where *C. achates* was present. Although these effects were not statistically significant, they could suggest an interaction whereby *C. achates* may impact *C. maculosa* populations when the plant is drought stressed. Carefully designed experiments are currently underway to evaluate this hypothesis by examining the independent and interacting effects of spring precipitation inputs and *C. achates* herbivory on *C. maculosa* populations (D. E. Pearson and Y. K Ortega, ongoing research). Although it is presently unclear whether current biocontrol agents actually offer sufficient control over *C. maculosa* to relieve indirect nontarget effects due to food-web interactions, the case study of *C. maculosa* biocontrol illustrates the importance of efficacy for safe and effective biocontrol. If biocontrol is successful, i.e., reduces populations of the target weed below its ecological threshold of impact, then indirect nontarget effects should dissipate to levels that are biologically insignificant (Pearson and Callaway 2003, 2005). As the *C. maculosa* saga continues to unfold, *C. maculosa* may ultimately provide an empirical case study showing how efficacy can guard against indirect nontarget effects, but such an outcome will depend on the degree to which biocontrol agents can actually be shown to reduce *C. maculosa* populations over large areas.

SECONDARY INVASION

As investigations into the efficacy of weed management have begun to expand beyond the target weed to incorporate general vegetation responses, an alarming trend is emerging. Successful control of the target species is far too often resulting in their replacement by secondary invaders (Table 1). This result does not appear to be limited to any one particular management strategy, target weed, or secondary invader. For example, secondary invaders flourish following *C. maculosa* control using herbicide, biocontrol, and grazing (Olson et al. 1997; Story et al. 2006; Ortega and Pearson in review, Rinella et al. in review). Control of crownvetch (*Coronilla varia*), St. Johnswort (*Hypericum perforatum*), artichoke thistle (*Cynara cardunculus*), and diffuse knapweed (*Centaurea diffusa*) have also given rise to secondary invaders (Huffaker and Kennett 1959; Symstad 2004; Bush et al. 2007; Seastedt et al. 2008). Secondary invasion following successful control of *H. perforatum* in the mid nineteen hundreds (Huffaker and Kennett 1959) indicates this is not merely a contemporary problem, but likely results any time additional species of strong invaders are present at a site where a dominant weed is controlled and the secondary invader proves relatively insensitive to the control method.

Although secondary invasion appears somewhat ubiquitous, a few general patterns do emerge. First and foremost, most prominent secondary invaders we could find documentation of were grasses. The principal grass was cheatgrass (*Bromus tectorum*), but Kentucky bluegrass (*Poa pratensis*) was another key secondary invader. Additionally, most target weeds were forbs. Thus, there is a general pattern suggesting that control of invasive forbs is shifting communities toward exotic grasses, though we also found one case where control of

Table 1. Secondary invasion following control of the target weed.

Target Weed	Secondary Invader	System	Control Method	Source
Artichoke thistle (*Cynara cardunculus*)	Exotic spp.	grassland	herbicide	Seastedt et al. (2008)
Diffuse knapweed (*Centaurea diffusa*)	Cheatgrass (*Bromus tectorum*)	grassland	biocontrol	Bush et al. (2007)
Leafy spurge (*Euphorbia esula*)	Canada thistle (*Circium arvense*)	meadow	herbicide	Smith et al. (2006)
Crownvetch (*Coronilla varia*)	Kentucky bluegrass (*Poa pratensis*)	prairie	herbicide	Symstad (2004)
Spotted knapweed (*Centaurea maculosa*)	Cheatgrass (*Bromus tectorum*)	grassland	herbicide	Ortega and Pearson in review
Spotted knapweed (*Centaurea maculosa*)	Cheatgrass (*Bromus tectorum*)	agricultural	biocontrol	Story et al. (2006)
Spotted knapweed (*Centaurea maculosa*)	Kentucky bluegrass (*Poa pratensis*) and other exotic grasses	grassland	herbicide	Tyser et al. (1998)
Spotted knapweed (*Centaurea maculosa*)	Kentucky bluegrass (*Poa pratensis*)	grassland	grazing	Olson et al. (1997)
Spotted knapweed (*Centaurea maculosa*) Sulfur cinquefoil (*Potentilla recta*)	Cheatgrass (*Bromus tectorum*) and other exotic grasses	grassland	herbicide	Sheley et al. (2006)
St. Johnswort (*Hypericum perforatum*)	Cheatgrass (*Bromus tectorum*) and other exotic grasses	grassland	biocontrol	Huffaker and Kennett (1959)

dominant exotic grasses allowed increases in exotic forbs (Cox and Allen 2008). To some extent, this pattern could be attributed to control methods such as herbicides and grazing that tend to generically suppress either forbs or grasses, thereby favoring weeds in unaffected functional groups. However, even controlling exotic weeds with highly selective methods such as biocontrol can give rise to secondary invaders like downy brome (a.k.a cheatgrass) (Huffaker and Kennett 1959; Story et al. 2006). This raises the question of whether, at least in some cases, weed control is simply accelerating general patterns of invader succession, or creating an "invasive species treadmill" (McEvoy and Coombs 2000). For example, seven years of monitoring of native grasslands, *C. maculosa*-dominated grasslands, and *C. maculosa*-dominated grasslands that had been treated with picloram showed that although cheatgrass invasion was greatly accelerated in herbicide treated areas, it also increased substantially in the native and untreated *C. maculosa*-invaded sites over time (Y. K. Ortega and D. E. Pearson, unpublished data). None of these sites were grazed by domestic herbivores over the study period, suggesting that disturbance was not increased over natural levels. Cheatgrass may simply be increasing in intermountain regions, and it may ultimately overtake *C. maculosa*-invaded habitats. In general, remarkably little is known about invader succession. However, knowledge of invader succession may prove extremely valuable for predicting and preventing secondary invasion following various weed management activities.

Recent ecological studies indicate that understanding the processes underlying invasion may be essential to understanding the problem of secondary invasion. If invaders are acting as passengers in a system and are merely responding to changes in disturbance regimes or other processes like climate change (Hobbs and Huenneke 1992; MacDougall and Turkington 2005), or if invasion persists as a residual effect of system-level changes such as loss of native seed inputs (Seabloom et al. 2003), then management must move beyond suppression of the dominant weed to management of system-level processes, where possible, to reverse the invasion trend. In contrast, if invaders are drivers, particularly as in the case of some grasses that usurp system-level processes such as fire to their advantage (D'Antonio and Vitousek 1992), the problem may be untenable and resources may be best deployed elsewhere. Additional ecological studies are needed that identify such system-level causes of invasion as well as studies that begin to apply these understandings to restoration problems (e.g., Orrock et al. 2008).

Unfortunately, any current discussions about secondary invasion are largely speculative, because there simply is not enough information presently available to effectively assess patterns and potential causes, at least not in a readily accessibly form. Our searches of databases for studies documenting secondary invasion proved largely fruitless, because there is no formally recognized term for this phenomenon and many studies only casually note increases in secondary exotic species following weed control. One reason for this may be that secondary invasion by grasses in many cases may not be recognized as a problem because exotic grasses, even cheatgrass, are commonly valued as forage species (Mosely 1996). Similarly, exotic grasses are rarely included on noxious weed lists (Rice 2008). We suggest that more attention should be paid to evaluating the response of exotic plants to weed control in natural areas, whether exotics are grasses or forbs. Formal use of terminology such as "secondary invasion" to describe this phenomenon and its inclusion in published key word lists could rapidly expand our knowledge of the extent of this problem through meta-analyses that may possibly suggest some solutions. Current information suggests that secondary invasion may pose a serious and widespread problem. In fact, it may well be the greatest challenge to successful weed management in natural areas, as natural areas are increasingly bombarded by hoards of aggressive invaders trying to gain dominance. There is currently a gaping hole in our understandings of the extent and causes of secondary invasion that represents a major impediment to weed management.

Conclusion

Natural areas are highly complex systems involving many individual species whose composition and relative abundance both influence and are influenced by a variety of community interactions and ecosystem-level processes. Exotic plant invasions affect not only the biomass and richness of native plants, but higher trophic levels and many of the processes that structure native communities and systems. Moreover, weed management tools themselves can interact with native species and system processes in complex ways, as they are rarely so selective that they affect only the target weed. Even when management tools are highly selective, directly affecting only the target weed, they can still have strong side effects on nontarget species due to complex community interactions. Thus, weed management in natural areas is in essence community management or directed community assembly. In order to effectively and predictably manage invaded communities, we need to understand how the weed affects species, community-level, and system-level processes. We also need to understand how prospective management tools interact with the weed and other system components and processes to determine the potential for management side effects. Only with this information can we effectively evaluate the potential for management actions to mitigate certain weed impacts and weigh these against potential management side effects to ensure that the ultimate management outcome is one that maximizes efficacy and minimizes deleterious side effects. Secondary invasion appears to be a large and increasing obstacle to successful weed management because it derails attempts to effectively reassemble a community even after successful control of the target weed. Better understandings of the processes underlying exotic plant invasions are needed to evaluate the causes of secondary invasions and the potential for addressing this serious problem. We urgently need more studies documenting community- and ecosystem-level outcomes of weed management measures in order evaluate the extent to which such actions mitigate invader impacts but also incur deleterious side effects and promote secondary weed invasions.

REFERENCES

Arnott, R., S. Finger, and C. U. M. Smith. 2003. Trepanation discovery, history, theory. Swets and Zeitlinger Publishers, Lisse, The Netherlands.

Bais, H. P., R. Vepachedu, S. Gilroy, R. M. Callaway, and J. M. Vivanco. 2003. Allelopathy and exotic plants: from genes to invasion. Science 301:1377–80.

Bricker, M. J. M. Maron, and D. E. Pearson. In press. Small mammal seed predation reduces forb recruitment and abundance in semi-arid grasslands. Ecology

Bush, R. T., T. R. Seastedt, and D. Buckner. 2007. Plant community response to the decline in diffuse knapweed in a Colorado grassland. Ecological Restoration 25:169-174.

Callaway, R. M., and W. M. Ridenour. 2004. Novel weapons: invasive success and the evolution of increased competitive ability. Frontiers and Ecology and the Environment 2:436–443.

Childs, J. E., T. G. Ksiazek, C. F. Spiropoulou, J. W. Krebs, S. Morzunov, G. O. Maupin, K. L. Gage, J. Sarinsky, R. Enscore, J. Frey, C. J. Peters, and S. T. Nichol. 1994. Serologic and genetic identification of *Peromyscus maniculatus* as the primary rodent reservoir for a new hantavirus in the southwestern United States. Journal of Infectious Diseases 169:1271-1280.

Coombs, E. M., J. K. Clark, G. L. Piper, and J. A. F. Cofrancesco. 2004. Biological control of invasive plants in the United States. Oregon State University Press, Corvallis, OR.

Cox, R. D., and E. B. Allen. 2008. Stability of exotic annual grasses following restoration efforts in southern California coastal sage scrub. Journal of Applied Ecology 45:495-504.

Crone, E. E., M. Marler, and D. E. Pearson. 2009. Non-target effects of broadleaf herbicide on native perennial forbs: a demographic framework for assessing and minimizing impacts. Journal of Applied Ecology, in press.

D'Antonio, C. M., and P. M. Vitousek. 1992. Biological invasions by exotic grasses, the grass/fire cycle, and global change. Annual Review of Ecology and Systematics 23:63-87.

D'Antonio, C. M., N. E. Jackson, C. C. Horvitz, and R. Hedberg. 2004. Invasive plants in wildland ecosystems: merging the study of invasion processes with management needs. Frontiers in Ecology and the Environment 2:513-521.

Davis, M. A., J. P. Grime, and K. Thompson. 2000. Fluctuating resources in plant communities: a general theory of invasibility. Journal of Ecology 88:528-534.

DiTomaso, J. M. 2000. Invasive weeds in rangelands: species, impacts, and management. Weed Science 48:255-265.

Dukes, J. S. 2002. Species composition and diversity affect grassland susceptibility and response to invasion. Ecological Applications 12:602-617.

Evans, W. E., and H. L. McLeod. 2003. Pharmacogenomics – drug disposition, drug targets, and side effects. New England Journal of Medicine 384:538-549.

Gouden, R. D., and S. M. Louda. 1976. Biotic interference with insects imported for weed control. Annual Review of Entomology 21:325-342.

Gurr, G., and S. Wratten. 2000. Biological control: measures of success. Kluwer Academic Publishers. Dordecht, The Netherlands.

Groh, H. 1944. Canadian Weed Survey. 2nd Annual Report. Canada Department of Agriculture.

Harris, P., and J. H. Myers. 1984. Centaurea diffusa Lam. and C. maculosa Lam. s. lat., diffuse and spotted knapweed (Compositae). Biological Control Programmes against Insects and Weeds in Canada 1969-1980. Chapter 31:127-137.

Henneman, M. L., and M. J. Memmott. 2001. Infiltration of a Hawaiian Community by Introduced Biological Control Agents. Science 293:1314-1316.

Hobbs, R. J. and L. F. Huenneke. 1992. Disturbance, diversity, and invasion: implications for conservation. Conservation Biology 6:324-337.

Hobbs, R. J., and S. E. Humphries. 1995. An integrated approach to the ecology and management of plant invasions. Conservation Biology 9:761-770.

Holt, R.D., and M. E. Hochberg. 2001. Indirect interactions, community modules and biological control: a theoretical perspective. In: Wajnberg, E., Scott, J.K., Quimby, P.C. (Eds.), Evaluating Indirect Effects of Biological Control. CABI Publishing. Wallingford, UK, pp. 13-38.

Howarth, F. G. 1991. Environmental impacts of classical biological control. Annual Reviews of Entomology 36:485-509.

Huffaker, C. B., and C. E. Kennett. 1959. A ten-year study of vegetational changes associated with biological control of Klamath weed. Journal of Range Management 12:69-82.

Hulme, P. E. 2006. Beyond control: wider implications for the management of biological invasions. Journal of Applied Ecology 43:835-847.

Lacey, J. R., C. B. Marlow, and J. R. Lane. 1989. Influence of spotted knapweed (*Centaurea maculosa*) on surface runoff and sediment yield. Weed Technology 3:627-631.

LeJeune, K. D., and T. R. Seastedt. 2001. *Centaurea* species: the forb that won the west. Conservation Biology 15:1568-1574.

Lenhoff, E. A., B. D. Maxwell, and L. J. Rew. In press. Quantifying invasiveness of plants: A test case with yellow toadflax (*Linaria vulgaris*). Invasive Plant Science & Management (in press).

Levine, J. M., M. Vila, C. M. D'Antonio, J. S. Dukes, K. Grigulis, and S. Lavorel. 2003. Mechanisms underlying the impacts of exotic plant invasions. Proceedings of the Royal Society of London 270:775-781.

Loeser, M. R. R., T. D. Sisk, and T. E. Crews. 2007. Impact of grazing intensity during drought in an Arizona grassland. Conservation Biology 21:87-97.

Louda, S. M., D. Kendall, J. Connor, and D. Simberloff. 1997. Ecological effects of an insect introduced for the biological control of weeds. Science 277:1088-1090.

MacDougall, A. S., and R. Turkington. 2005. Are invasive species the drivers or passengers of ecological change in highly disturbed plant communities? Ecology 86:42-55.

Mack, R. N., D. Simberloff, W. M. Lonsdale, H. Evans, M. Clout, and F. A. Bazzaz. 2000. Biotic invasions: causes, epidemiology, global consequences, and control. Ecological Applications 10:689-710.

Maron, J. L., and M. J. Kauffman. 2006. Habitat-specific impacts of multiple consumers on plant population dynamics. Ecology 87:113–124.

Maron, J. L., and M. J. Marler. 2007. Native plant diversity resists invasion at both low and high resource levels. Ecology 88:2651-2661.

Maxwell, B. D., E. A. Lehnhoff, and L. J. Rew. In press. The rationale for monitoring invasive plant populations as a crucial step for management. Invasive Plant Science & Management (In press).

McEvoy, P. B. 1996. Host specificity and biological pest control. BioScience 46:401-405.

McEvoy, P. B., and E. M. Coombs. 2000. Why things bite back: unintended consequences of biological weed control. Pages 167-194 *in* P. A. Follet and J. J. Duan, editors. Nontarget effects of biological control. Kluwer Academic Publishers, Boston, Massachusetts, USA.

Mosley, J. C. 1996. Prescribed sheep grazing to suppress cheatgrass: a review. *Sheep and Goat Research Journal*. 12: 74–81.

Mueggler, W. F., and W. L. Stewart. 1980. Grassland and shrubland habitat types of western Montana. USDA Forest Service, General Technical Report INT-66, Ogden, Utah, USA.

Müller-Shärer, H., and U. Schaffner. 2008. Classical biological control: exploiting enemy escape to manage plant invasions. Biological Invasions. 10:859-874.

Niggemann, B., and C. Gruber. 2003. Side-effects of complementary and alternative medicine. Allergy 58:707-716.

Olson, B. E., R. T. Wallander, and J. R. Lacey. 1997. Effects of sheep grazing on a spotted knapweed-infested Idaho fescue community. Journal of Range Management 50:386-390.

Orrock J. L., M. S. Witter, and O. J. Reichman. 2008. Native consumers and seed limitation constrain the restoration of a native perennial grass in exotic habitats. Restoration Ecology 17:148-157.

Ortega, Y. K., D. E. Pearson, and K. S. McKelvey. 2004. Effects of introduced biological control agents and exotic plant invasion on native deer mouse populations. Ecological Applications 14:241-253.

Ortega, Y. K., and D. E. Pearson. 2005. Strong versus weak invaders of natural plant communities: assessing invasibility and impact. Ecological Applications 15:651-661.

Ortega Y. K., and D. E. Pearson. In review. Effects of a common broadleaf herbicide treatment on grassland community dominants vary with initial levels of exotic plant invasion.

Ortega, Y. K., K. S. McKelvey, and D. L. Six. 2006. Invasion of an exotic forb impacts reproductive success and site fidelity of a migratory songbird. Oecologia 149:340-351.

Pearson, D. E. 2009. Invasive plant architecture alters trophic interactions by changing predator abundance and behavior. Oecologia 159:549-558.

Pearson, D. E., K. S. McKelvey, and L. F. Ruggiero. 2000. Non-target effects of an introduced biological control agent on deer mouse ecology. Oecologia 122:121-128.

Pearson, D. E., and R. M. Callaway. 2003. Indirect effects of host-specific biological control agents. Trends in Ecology and Evolution 18: 456-461.

Pearson, D. E., and R. M. Callaway. 2005. Indirect nontarget effects of host-specific biological control agents: implications for biological control. Biological Control 35:288-298.

Pearson, D. E., and R. M. Callaway. 2006. Biological control agents elevate hantavirus by subsidizing deer mouse populations. Ecology Letters 9: 443–450.

Pearson, D. E., and R. J. Fletcher, Jr. 2008. Mitigating exotic impacts: restoring deer mouse populations elevated by an exotic food subsidy. Ecological Applications 18:321-334.

Pearson, D. E,. and R. M. Callaway. 2008. Weed biocontrol insects reduce native plant recruitment through second-order apparent competition. *Ecological Applications* 18:

Pemberton, R. W. 2000. Predictable risk to native plants in weed biological control. Oecologia 125:489-494.

Pimentel, D., R. Zuniga, and D. Morrison. 2005. Update on the environmental and economic costs associated with alien-invasive species in the United States. Ecological Economics 52:273-288.

Pokorny, M. L., R. L. Sheley, T. J. Svejcar, and R. E. Engel. 2004. Plant species diversity in a grassland plant community: evidence for forbs as a critical management. Western North American Naturalist 64:219-230.

Pokorny, M. L., R. L. Sheley, C. A. Zabinski, R. E. Engel, T. J. Svejcar, and J. J. Borkowski. 2005. Plant functional group diversity as a mechanism for invasion resistance. Restoration Ecology 13:448-459.

Potts, S. G., B. Vulliamy, A. Dafani, G. Ne'eman, and P. Willmer. 2003. Linking bees and flowers: how do floral communities structure pollinator communities? Ecology 84:2628-2642.

Pysek, P., D. M. Richardson, J. Pergl, V. Jarosik, Z. Sixtova, and E. Weber. 2008. Geographic and taxonomic biases in invasion ecology. Trends in Ecology and Evolution 23:237-244.

Randall, J. M. 1996. Weed control for the preservation of biological diversity. 1996. Weed Technology 10:370-383.

Rew, L.J., E.A. Lehnhoff and B.D. Maxwell. 2007. Non-indigenous species management using a population prioritization framework. Can Journal of Plant Science 87:1029-1036.

Rice, P. M., J. C. Toney, D. J. Bedunah, and C. E. Carlson. 1997a. Plant community diversity and growth form responses to herbicide applications for control of *Centaurea maculosa*. Journal of Applied Ecology 34:1397-1412.

Rice, P. M., J. C. Toney, D. J. Bedunah, and C. E. Carlson. 1997b. Elk winter forage enhancement by herbicide control of spotted knapweed. Wildlife Society Bulletin 25:627-633.

Rice, P. M., and J. C. Toney. 1998. Exotic weed control treatments for conservation of fescue grassland in Montana. Biological Conservation 85:83-95.

Rice, P. M. 2008. INVADERS Database System (http://invader.dbs.umt.edu). Division of Biological Sciences, University of Montana, Missoula, Montana.

Ridenour, W. M., and R. M. Callaway. 2001. The relative importance of allelopathy in interference: the effects of an invasive weed on a native bunchgrass. Oecologia 126:444-450.

Rodriguez, L. F. 2006. Can invasive specs facilitate native species? Evidence of how, when, and why these impacts occur. Biological Invasions 8:927-939.

Seabloom, E. W., E. T. Borer, V. L. Boucher, R. S. Burton, K. L. Cottingham, L. Goldwasser, W. K. Gram, B. E. Kendall, and F. Micheli. 2003. Competition, seed limitation, disturbance, and reestablishment, of California native annual forbs. Ecological Applications 13:575-592.

Seastedt, T. R., R. J. Hobbs, and K. N. Suding. 2008. Management of novel ecosystems: are novel approaches required? Frontiers in Ecology and the Environment (In press)..

Shea, K., D. Kelly, A. W. Sheppard, and T. L. Woodburn. 2005. Context-dependent biological control of an invasive thistle. Ecology 86: 3174–3181.

Sheley, R. L., and J. S. Jacobs. 1997. Repsonse of spotted knapweed and grass to picloram and fertilizer combinations. Journal of Range Management 50:263-267.

Sheley, R., T. J. Svejcar, and B. D. Maxwell. 1996. A theoretical framework for developing successional weed management strategies on rangeland. Weed Technology 10:766-773.

Sheley, R. L., J. S. Jacobs, and M. F. Carpinelli. 1998. Distribution, biology, and management of diffuse knapweed (*Centaurea diffusa*) and spotted knapweed (*Centaurea maculosa*). Weed Technology 12:353-362.

Sheley, R. L., J. M. Mangold, and J. L. Anderson. 2006. Potential for successional theory to guide restoration of invasive-plant dominated rangeland. Ecological Monographs 76:365-379.

Simberloff, D., and P. Stiling. 1996. How risky is biological control? Ecology 77:1965-1974.

Smith, R. G., B. D. Maxwell, F. D. Menalled, and L. J. Rew. 2006. Lessons from agriculture may improve the management of invasive plants in wildland systems. Frontiers in Ecology and the Environment 4:428-434.

Story, J. M., K. W. Boggs, W. R. Good, L. J. White, and R. M. Nowierski. 1995. Cause and extent of predation on *Urophora* spp. larvae (Diptera: Tephritidae) in spotted knapweed capitula. Environmental Entomology 24:1467-1472.

Story, J. M., N. W. Callan, J. G. Corn, and L. J. White. 2006. Decline of spotted knapweed density at two sites in western Montana with large populations of the introduced root weevil, *Cyphocleonus achates* (Fahraeus). Biological Control 38:227–232.

Story, J. M., L. Smith, J. G. Corn, and L. J. White. 2008. Influence of seed head–attacking biological control agents on spotted knapweed reproductive potential in western Montana over a 30-year period. Environmental Entomology 37:510-519.

Strange, R. M., K. M. Lindsay, and R. S. Price. 1979. Knapweeds: British Columbia's undesirable aliens. Rangelands 1:141-143.

Sturdevant, N., S. Kegley, Y. Ortega, and D. Pearson. 2006. Evaluation of establishment of *Cyphocleonus achates* and its potential impact on spotted knapweed. USDA FS FHP General Technical Report 06-08:1-9.

Symstad, A. J. 2004. Secondary invasion following the reduction of *Coronilla varia* (Crownvetch) in Sand Prairie. American Midland Naturalist 152:183-189.

Thomas, M. B., and A. M. Reid. 2007. Are exotic natural enemies an effective way of controlling invasive plants? Trends in Ecology and Evolution 22:447-453.

Thompson, M. J. 1996. Winter foraging response of elk to spotted knapweed removal. Northwest Science 70: 10-19.

Trammell, M. A., and J. L. Butler. 1995. Effects of exotic plants on native ungulate use of habitat. Journal of Wildlife Management 59:808-816.

Tu, M., C. Hurd, and J. M. Randall. 2001. Weed Control Methods Handbook: Tools and Techniques for Use in Natural Areas. The Nature Conservancy, http://tncweeds.ucdavis.edu, Version: April 2001.

Tyser, R. W., J. M. Asebrook, R. W. Potter, and L. L. Kurth. 1998. Roadside revegetation in Glacier National Park, USA: effects of herbicide and seeding treatments. Restoration Ecology 6: 197-206.

Tyser, R. W., and C. H. Key. 1988. Spotted knapweed in natural area fescue grasslands: an ecological assessment. Northwest Science 62:151–160.

Van Driesche, R., M. Hoddle, and T. Center. 2008. Control of pests and weeds by natural enemies. Blackwell Publishing, Malden, Massachusettes, USA.

Vitousek, P. M. 1990. Biological invasions and ecosystem processes: towards an integration of population biology and ecosystem studies. Oikos 57:7-13.

Walker, J. W., S. L. Kronberg, S. L. Al-Rowaily, and N. E. West. 1994. Comparison of sheep and goat preferences for leafy spurge. Journal of Range Management 47:429-434.

White, E. M., J. C. Wilson, and A. R. Clarke. 2006. Biotic indirect effects: a neglected concept in invasion biology. Biodiversity Research 12:443-455.

Wright, A. L., and R. G. Kelsey. 1997. Effects of spotted knapweed on a cervid winter-spring range in Idaho. Journal of Range Management 50:487-496.

Zavaleta, E. S., and K. B. Hulvey. 2004. Realistic species losses disproportionately reduce grassland resistance to bological invaders. Science 306:1175-1177.

In: Weeds: Management, Economic Impacts and Biology
Editor: Rudolph V. Kingely
ISBN: 978-1-60741-010-2

Chapter 2

OPPORTUNITIES FOR ECOLOGICALLY-BASED AEROBIC RICE CROPPING SYSTEMS: WEED COMPETITIVENESS OF GENOTYPES

***D.L. Zhao*[a]*, G.N. Atlin*[b]*, L. Bastiaans*[c] *and J.H.J. Spiertz*[c]**

[a] Plant Breeding, Genetics and Biotechnology, International Rice Research Institute (IRRI), DAPO Box 7777, Metro Manila, Philippines

[b] CIMMYT, Apdo. Postal 6-641 06600 Mexico, D.F., Mexico

[c] Crop and Weed Ecology, Plant Sciences, Wageningen University, P.O. Box 430, 6700AK Wageningen, The Netherlands

ABSTRACT

Increasing water scarcity in Asia is enforcing the adoption of water-saving rice systems such as aerobic rice, the production of direct-seeded rice in fields which are not puddled or flooded. Aerobic rice systems can greatly reduce the water requirement for rice production, but the lack of a standing water layer to suppress weed germination and growth can result in severe weed infestation. Concerns about the environmental consequences of herbicide use and the high labor costs of mechanical weeding highlight the importance of weed-competitive cultivars in the development of integrated management systems for aerobic rice. To identify crop traits related to weed-competitiveness (WC), to develop a selection protocol for WC, and to assess the role of WC in weed management, field experiments were conducted with aerobic and upland rice genotypes during 2001-2004 on the farm of the International Rice Research Institute (IRRI) in the Philippines. The research revealed that weed-suppressive genotypes are an effective component of weed management, especially when grown at an appropriate seeding rate (300 viable seeds m^{-2}). Improving weed competitiveness in aerobic rice is feasible because there exists a large genetic variability in weed-suppressive ability (WSA) among genotypes or germplasm, and because yielding ability and WSA are compatible. Fast early growth rather than plant type (erect or droopy) is the main determinant of the WSA. Indirect selection for rice yield under weed-free conditions, combined with selection for early crop vigor can simultaneously improve WSA and crop yield under weed competition.

Key words: Indirect selection index; plant type; vegetative growth; weed-suppressive ability; yielding ability

Abbreviations:
GC: canopy ground cover;
H: broad-sense heritability;
ISE: indirect selection efficiency;
LAI: leaf area index;
WAS: weeks after sowing;
WC: weed competitiveness;
WSA: weed-suppressive ability;
WT: weed tolerance.

INTRODUCTION

In irrigated lowland rice systems, where 75% of world rice is currently produced, large regions will be hampered by water scarcity by 2025 (Tuong and Bouman, 2003). To ensure rice production with less water, two promising water-saving rice systems, alternate wetting and drying and aerobic rice, need to be further developed. Aerobic rice is direct seeded in non-puddled, non-flooded fields, and is ideally grown in soils where water content is maintained at or a little below field capacity through need-based irrigation, or in favorable rainfed environments. Potentially, it may produce moderately high yields (6 – 7 Mg ha^{-1}) while reducing water use by more than 50% compared with conventional lowland irrigated rice (Wang et al., 2002). Actual yields are usually much lower (3 – 4 Mg ha^{-1}), because of management problems, including occasional water stress and poor weed control. Both aerobic and upland rice grow under aerobic conditions; however, the latter is less input-responsive, and completely dependent on rainfall, thus produces low yields (1.0 – 1.5 Mg ha^{-1}, Anonym, 1993).

In direct seeded aerobic and upland rice systems where rice and weeds germinate simultaneously, the lack of a 'head start' of rice seedlings over weeds and the absence of a water layer that suppresses weeds, unlike in transplanted lowland rice systems, result in more severe weed infestation than that in lowland rice (De Datta and Llagas, 1984). Weeds in aerobic rice may cause a yield loss from 30 to 100%, and thus are the greatest constraint to aerobic rice production (De Datta and Llagas, 1984; Oerke and Dehne, 2004). Chemical weed control is effective (Singh et al., 2008); however, the intensive use of herbicides results in environmental pollution and selection for herbicide-resistant weed biotypes (Heap, 2006). Weeding by hand or with simple tools is labor-intensive (Roder and Keohulapha, 1997), and is often not properly done due to high cost or unavailability of labor (Johnson, 1996). A more environment-friendly and less labor-intensive weed control method is needed to enhance aerobic rice adoption while protecting the environment. Weed-competitive genotypes may be an effective tool in integrated weed management (Fischer et al., 2001; Lemerle et al., 2001[b]).

Competition occurs in communities when two or more plants seek common resources, such as mineral nutrients, light, and water, within a common space. It is called 'intraspecific competition' if the competition occurs between individuals of the same species, and 'interspecific competition' if between individuals of different species (Lemerle et al., 2001[b]).

A crop plant experiences intraspecific competition from its neighbor crop plants in a weed-free field, but experiences both intra- and interspecific competition from its neighbor crop and weed plants in a weedy field. Weed competitiveness (WC) is the ability of a crop to compete against weeds for light, nutrients and water. WC has two components: weed tolerance (WT), the ability to maintain high yields despite weed competition, and weed-suppressive ability (WSA), the ability to reduce weed growth through competition (Jannink et al., 2000). Differences in WSA among cultivars can be directly determined by assessing weed biomass or weed seed production in plots under weed competition; the higher the weed biomass or weed seed production the weaker the WSA. Differences in WT can only be compared in terms of crop grain yield under weed competition among cultivars with the same yield potential and WSA (Gibson and Fischer, 2004). Jannink et al. (2000) and Jordan (1993) advocate breeding for WSA over WT because suppressing weeds reduces weed seed production and benefits weed management in the future while tolerating weeds only benefits the current growing season. Weed pressure from unsuppressed weeds increases the likelihood of crop yield loss, irrespective of the crop's tolerance. In this research, WSA was focussed on.

Substantial genetic variation in WC has been found in upland or aerobic (Garrity et al., 1992) and lowland rice (Fischer et al., 1997; Gibson et al., 2003), between *indica* and *japonica* subspecies (Oka, 1960), and between *O. glaberrima* and *O. sativa* species (Dingkuhn et al., 1999; Fofana & Rauber 2000), suggesting that differences in WC often exist among cultivated rice species and subspecies. Asian cultivated rice (*O. sativa*) was classified using isozymes by Glaszmann (1987) into *indica, aus, ashina, rayada, aromatic* and *japonica* varietal groups. More recently, it was differentiated by Garris et al. (2005), using simple sequence repeats and chloroplast sequences, into *indica, aus, aromatic*, temperate *japonica*, and tropical *japonica* groups. Knowledge on WC is very limited for these groups, especially for *indica*, tropical *japonica, aus* and their progenies, which are extensively used in tropical aerobic rice breeding programmes. Knowledge about their WC is needed to design breeding programs aimed at aerobic systems.

Weed competitiveness is often linked to plant height (Garrity et al., 1992), tiller number (Fischer et al., 1997), early height growth rate (Caton et al., 2003), early crop biomass (Ni et al., 2000), leaf area index (LAI) (Dingkuhn et al., 1999), specific leaf area (Audebert et al., 1999), and canopy ground cover (GC) (Lotz et al., 1995). However, despite many years of research and considerable evidence of varietal differences in WC, there have been limited efforts to breed for improved WC (Zimdahl, 2004). Gibson and Fischer (2004) attribute the limited progress in breeding for WC to: (1) the successful chemical control of weeds, which has led researchers to focus on herbicide use with less emphasis on other control methods and (2) the tradeoff between yield and WC suggested by earlier researchers (Jennings and Jesus, 1968; Kawano et al., 1974). Recent research, however, has shown that WC and yield potential can be compatible (Garrity et al., 1992; Ni et al., 2000; Fischer et al., 2001; Gibson et al., 2003).

Selection for WC may be done directly in the presence of weeds, or indirectly, under non-competitive conditions for secondary traits related to WC. Direct selection for WC entails growing each genotype in the presence of weeds to measure weed biomass and crop yield as selection criteria (Wall, 1983). High labour requirements and high residual variances of these measurements make direct selection impractical for most breeding programmes. Indirect selection, which can be carried out in the absence of weeds and may permit selection to be started earlier in a breeding programme, is likely to be easier and less expensive.

Indirect selection efficiency (ISE) of a trait as a selection criterion is a function of the broad-sense heritability (H) of the selection criterion under weed-free conditions and of the target trait under weed competition, and their genetic correlation (Falconer, 1989). Traits that are potentially useful indirect selection criteria for WC should be heritable under weed-free conditions and highly correlated with both weed biomass and crop yield under competition. They should also be practical for use in large breeding populations to achieve adequate selection intensity (Atlin et al., 2001). Indirect selection under weed-free conditions is preferable to direct selection when ISE is close to or greater than 1, and indirect selection is less expensive than direct selection. There are few reports in the literature of the heritability of WC and its component traits. Fischer et al. (1995, 1997) and Haefele et al. (2004) reported that some vegetative traits measured in weed-free rice variety trials, including LAI and tiller number, were uncorrelated with weed growth or competition-induced yield loss, and suggested that only direct selection for WC would be effective. However, the work of Jannink et al. (2000) on seedling height of soybean, Ni et al. (2000) on rice seedling biomass, and Gibson et al. (2003) on rice LAI and root growth during the vegetative stages suggests that some seedling traits measured in weed-free conditions are highly correlated with weed growth. Therefore, indirect selection for WSA, an important component of WC, may be feasible. However, destructively measured traits may be impractical selection criteria because of time, seed or land required when hundreds or thousands of lines must be assessed. If an easily-used selection protocol can be developed based on nondestructive measurements or ratings, breeding protocols for WC will be greatly simplified.

Increased seeding rates in lowland rice strengthen the ability of crops to suppress weeds while increasing crop yield under weedy conditions (Ni et al., 2004; Phuong et al., 2005). The combination of a competitive genotype and an optimal seeding rate may substantially improve weed suppression of crop, thus decreasing weeding operations. Genotype x seeding rate optimization for weed competitiveness is, however, rarely studied especially in aerobic rice systems.

The experiments described in this review of research conducted at the International Rice Research Institute (IRRI) on WC of aerobic rice aimed to (1) examine the magnitude of genotype and germplasm variability in WSA, (2) explore the feasibility of breeding for WC in aerobic rice, (3) identify traits that can be measured in weed-free environments, are heritable, and are highly correlated with both WSA and yield under weed competition, (4) develop indirect selection protocols for breeding for WC, and (5) test the efficacy of the combination of weed-competitive genotype with seeding rate in weed management.

MATERIALS AND METHODS

Experiments

Three experiments were conducted under aerobic conditions at the experimental station of IRRI in the Philippines:

Experiment 1. Forty aerobic and upland genotypes (*O. sativa*), consisting of 3 *aus*, 7 *indica*, and 23 *japonica* lines, and 7 lines derived from *indica*/*japonica* crosses, were grown in plots sized 4.5 m^2 under weed-free (no weeds were allowed to grow in an entire season) and weedy (one hand-weeding at 3 weeks after sowing (WAS), but weeds allowed to grow thereafter) conditions, respectively, arranged in an α-lattice design in each case, in the wet seasons of 2001−2003. Yield, yield components (harvest index and final crop biomass), weed biomass (weedy conditions only) at 13 WAS, and crop vigor (visually rated crop biomass with 1 as the least and 9 as the greatest vigor score) at 2 WAS were investigated in each season. In 2003, more detailed vegetative traits, including crop vigor (at 4 and 6 WAS), seedling height (at 3, 4, 6, and 9 WAS), plant erectness (visually rated with 1 as the most droopy and 9 as the most erect, at 3 and 6 WAS), crop biomass (at 4 and 9 WAS) and crop ground cover (green area as a percentage of total area in a photograph taken with a digital camera at 6 WAS) were studied.

Experiment 2. To examine the relationship between tiller development and competitiveness, the same set of 40 genotypes was grown in one row each in an α-lattice design under weed-free conditions. Plants in rows were thinned to one plant per hill spaced 5 cm apart. Tiller number per plant was measured twice (at 4 and 8 WAS) to relate vegetative tillering ability to WC.

Experiment 3. Three aerobic rice genotypes, Apo, IR60080-46A and IRAT216, that differ in yield potential and WSA, were grown at three seeding rates (100, 300 and 500 viable seeds m^{-2}) with or without seed priming under two weed management regimes (weed-free and weedy) in the 2003 wet season and 2004 dry season. The experiment was a split-plot design with weed management as main plots, and genotype × seeding rate or genotype × seeding rate × seed priming treatment as sub-plots. In this experiment, emergence, yield, weed biomass (at 8 and 12 WAS), and vegetative traits including tiller number, crop biomass and LAI at 2, 5 and 8 WAS were investigated.

ANALYSIS

The analyses were conducted using SAS (SAS Institute Inc., 1999-2001). Phenotypic correlations among traits were calculated on the basis of cultivar means over years, or over replicates within a year in some cases. Predicted broad-sense heritability (after Nyquist, 1991), genetic correlation (after Bernardo, 2002 and Cooper et al., 1996), and ISE (after Falconer, 1989) for traits were estimated (see details in Zhao, 2006[d]).

RESULTS AND DISCUSSION

Genetic Variability in Weed-Suppressive Ability

Substantial genetic variability in WSA was observed among the 40 genotypes tested under weed pressure levels (expressed as weed biomass averaged over all weedy plots within a year) of 305 g m^{-2} in 2001, 172 g m^{-2} in 2002, and 73 g m^{-2} in 2003. Mean weed biomass over three years for the genotypes ranged from 126 to 296 g m^{-2}, indicating that genotypes significantly differ in WSA (plant type differences associated with these differences in WSA are illustrated in Figure 1), and that there is room to genetically improve WSA in aerobic rice through breeding. The four germplasm groups differed in WSA too (Figure 2). Both *aus* and *indica* appeared to be more weed-suppressive than tropical *japonica* and the progenies of *indica*/*japonica* crosses, and the *indica* lines appeared to be more productive than the other germplasm. These results indicate that gains in WSA from breeding efforts can be expected, and that *aus* and *indica* can best be used as gene donors for WSA in the tropics. Genetic variability in WSA among aerobic or upland rice cultivars (Garrity et al., 1992), and similar germplasm differences (Oka, 1960) have been previously observed.

Figure 1. Canopy geometry of a less weed-competitive (left) vs. more weed-competitive (right) cultivar

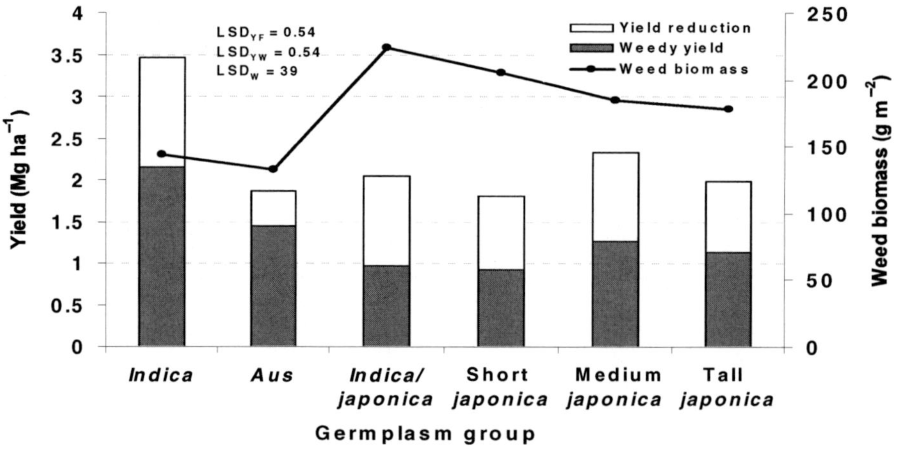

Figure 2. Weed-free yield (the whole bar) and weedy yield (the shaded bar) of crop, and weed biomass (the curve) for the six germplasm groups evaluated over three wet seasons of 2001–2003 at IRRI, Los Baños, Philippines. LSD_{YF}, LSD_{YW} and LS_{DW} are least significant differences at $P < 0.05$ for comparisons among groups in weed-free yield, weedy yield and weed biomass respectively (Zhao et al., 2006[c]).

COMPATIBILITY OF YIELDING ABILITY WITH WEED-SUPPRESSIVE ABILITY

Breeding for WC should not result in a trade-off for yield, because farmers are unlikely to adopt weed-competitive but low-yielding cultivars. Therefore, compatibility of yielding ability with WSA is a prerequisite for breeding weed-competitive aerobic rice. In this research, high-yielding ability with and without weed competition was found to be associated with low weed biomass (r = -0.84 and -0.67, respectively) (Figure 3), indicating that high-yielding ability and strong WSA are compatible, and may be combined in one genotype. This finding is in agreement with studies with lowland (Ni et al., 2000; Gibson et al., 2003) and upland rice (Garrity et al., 1992; Fofana and Rauber, 2000), and suggests that the trade-off between yield potential and WSA suggested by earlier researchers (Jennings and Aquino, 1968; Jennings and Jesus, 1968; Jennings and Herrera, 1968; Kawano et al., 1974) may not exist in aerobic rice.

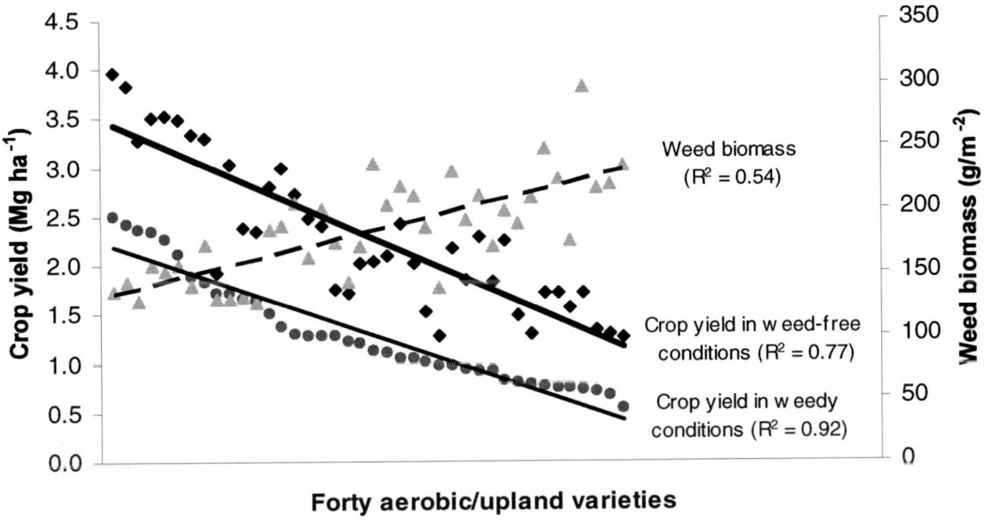

Figure 3. Yield of forty aerobic/upland rice varieties with and without weed competition, and the accompanying weed biomass averaged over three years of experimentation (Experiment 1; IRRI farm, wet season 2001-2003). Varieties are presented in descending order of crop yield under weedy conditions. (Source: Zhao et al., 2006[a])

CONSISTENCY OF CROP PERFORMANCE ACROSS WEED MANAGEMENT REGIMES AND YEARS

No weed management × genotype (or germplasm group) interaction was found in any experiment for almost all the studied traits (except harvest index, with a significant but small F value), including harvest traits, like grain yield and final crop biomass, and vegetative traits, like early crop vigor, early plant height, height growth rate, vegetative crop biomass, tiller number, LAI, plant erectness, and GC. This result indicates that performance of a genotype at any growing stage is relatively consistent across weed management regimes. Traits expressed under weed-free and weedy environments were found to be positively correlated, further confirming the above observation. This finding implies that selection under weed-free conditions will result in a corresponding response under weedy conditions. In agreement with this result, Gibson et al. (2003) also found a lack of weed management × genotype interaction for yield in lowland rice. Lemerle et al. (2001[a]) reported a positive correlation between weed-free and weedy yield in wheat, and Caton et al. (2003) a close correlation for vegetative traits of aerobic rice under intra- and interspecific competition. Our findings and the results reported in the literature suggest that intra- and interspecific competition may differ in degree but not in kind. Therefore, cultivars performing better under weed-free conditions (intraspecific competition) are likely to perform relatively better under weedy conditions (both intra- and interspecific competition). This conclusion is supported by Goldberg and Landa (1991), who found that suppressive ability of a species does not change with changes in its surrounding species in natural environments. However, Fischer et al. (1997, 2001) found significant weed management × genotype interactions for plant height, crop biomass, and yield in lowland and upland rice.

Experiment 1, with 40 genotypes evaluated under weedy and weed-free conditions over three years, showed that variances for genotype × year interaction for traits including yield, final crop biomass, weed biomass and early vigor were all smaller than those for genotype, indicating that genotype performance is also relatively consistent across years. Similarly, Experiment 3, with three contrasting genotypes evaluated over two years, showed relative consistency in all the studied traits including yield, weed biomass, LAI, tiller number and vegetative crop biomass. These results from different multiyear experiments indicate that the relative yield potential and WSA of genotypes do not change greatly from year to year. This finding is supported by studies with upland (Garrity et al. 1992; Fischer et al., 2001) and lowland rice (Fischer et al., 1997; Gibson et al., 2003). Relatively consistent WSA and yielding ability under weed competition across years within a target environment is very important for breeding for WC. If relative genotype performance changes markedly from year to year, or place to place (or, in other words, if heritability is low), breeding efforts for these traits will be of little value.

HERITABILITY FOR YIELD, WEED-SUPPRESSIVE ABILITY AND THEIR RELATED WEED-FREE TRAITS

The estimated H for yield under weed competition and weed biomass, the two target traits in breeding for WC, amounted to 0.79 and 0.64 on a three-year basis, and 0.55 and 0.38 on a single-year basis, respectively, in Experiment 1 (Table 1). The magnitude of the estimates appears to permit reasonable gains from selection. The estimated H for weed-free yield and weed-free crop vigor at 2 WAS were 0.87 and 0.65 on a three-year basis, and 0.68 and 0.38 on a single-year basis, and 0.96 and 0.81 estimated on the basis of means from one weed-free trial in 2003, respectively. The other weed-free traits including crop vigor at 4 and 6 WAS, and plant height at 4 WAS, had H values all greater than 0.80 estimated on the 2003 weed-free trials. The magnitude of the H values for all these weed-free traits indicates that they may serve as indirect selection criteria if they are genetically correlated with the target traits. Heritability for WC-related traits in rice is rarely studied. A study with soybean by Jannink et al. (2000) showed that early plant height and WSA are both heritable.

Table 1. Broad-sense heritability (H)† of weed-free traits evaluated under weed-free conditions, and of weedy traits evaluated under weed competition, indirect selection efficiency (ISE) of the weed-free traits, and phenotypic (rP) and genetic (rG) correlations between the weed-free and weedy traits.

Trait	H	$Yield_{Weedy}$‡ (0.55±0.06)			Weed (0.38±0.10)		
		r_P	r_G	ISE	r_P	r_G	ISE
Vigor2§	0.81±0.05	0.54**	*0.67*	0.81	-0.52**	*-0.72*	1.06
Vigor4	0.88±0.03	0.75**	*0.91*	1.14	-0.68**	*-0.9*	1.38
Vigor6	0.83±0.05	0.67**	*0.83*	1.02	-0.63**	*-0.86*	1.27
Erectness3	0.92±0.02	0.60**	*0.71*	0.91	-0.52**	*-0.68*	1.06
Erectness6	0.93±0.02	0.57**	*0.67*	0.87	-0.30^{ns}	*-0.39*	0.62
Ground cover6	0.67±0.09	0.55**	*0.75*	0.83	-0.53**	*-0.81*	1.07
Height3	0.79±0.06	0.40*	*0.51*	0.61	-0.56**	*-0.78*	1.13
Height4	0.81±0.05	0.47**	*0.59*	0.71	-0.66**	*-0.91*	1.34
Height6	0.90±0.03	-0.09^{ns}	*-0.10*	0.13	-0.13^{ns}	*-0.17*	0.26
Height9	0.79±0.06	-0.02^{ns}	*-0.02*	0.03	-0.27^{ns}	*-0.38*	0.55
Height final	0.83±0.05	0.03^{ns}	*0.04*	0.04	-0.24^{ns}	*-0.32*	0.48
Biomass4	0.69±0.09	0.67**	*0.90*	1.01	-0.54**	*-0.82*	1.10
Biomass9	0.63±0.11	0.51**	*0.73*	0.78	-0.51**	*-0.81*	1.04
Biomass final	0.62±0.11	0.68**	*0.97*	1.02	-0.33*	*-0.52*	0.67
Tiller4	0.89±0.03	0.45**	*0.54*	0.68	-0.41**	*-0.55*	0.84
Tiller8	0.95±0.02	0.74**	*0.86*	1.13	-0.58**	*-0.74*	1.18
Yield	0.96±0.01	0.92**	*1.00*	1.40	-0.55**	*-0.69*	1.11

*, **, and ns indicate that phenotypic correlations differ significantly from 0 at $P<0.05$, $P<0.01$ and $P>0.05$, respectively;

† Heritability and standard error:

- H values in parenthesis in the first row are for a single year of testing and predicted using variance components estimated from weedy trials over three years (2001-2003);
- H values in the second column were estimated using single-year data from experiment 1 and 2 in 2003;

‡ $Yield_{Weedy}$ and Weed indicate yield under weed competition, and weed biomass, respectively;

§ Number following a weed-free trait in the first column indicates the number of weeks after seeding at which the trait was evaluated.

(Source: Zhao et al., 2006[b]).

Association of Yielding Ability Under Competition and Weed-Suppressive Ability with Weed-Free Traits

Among the weed-free traits studied in the three experiments, yield, crop vigor (at 2 – 6 WAS), early crop biomass (at ≤ 9 WAS), early plant height (at ≤ 4 WAS), and early tillering (at 4 and 8 WAS) were all phenotypically and/or genetically correlated positively with grain yield under weed competition but negatively with weed biomass (Table 1). These results indicate: (1) fast early growth of a crop is an attribute of strongly weed-suppressive genotypes; (2) weed-free vegetative traits of a crop are predictive of WSA and yielding ability under weed competition. A number of studies support these findings and link rapid early growth to WSA (Lemerle et al., 1996; Bastiaans et al., 1997; Dingkuhn et al., 1999; Ni et al., 2000; Jannink et al., 2000; Fofana and Rauber, 2000; Gibson et al., 2001, 2003; Caton et al., 2003;). However, a few studies showed a weak association between WSA and early height (Fischer et al., 2001), or WSA and early tillering (Garrity et al., 1992) in upland rice.

INDIRECT SELECTION FOR YIELDING ABILITY UNDER COMPETITION AND WEED-SUPPRESSIVE ABILITY

Indirect Selection Efficiency

Among all the investigated traits under weed-free conditions, early crop vigor and yield were found to have high estimated ISE for both crop yield under weed competition and weed biomass. On a multi-year basis, the ISE of crop vigor at 2 WAS was 0.80 for crop yield and 0.89 for weed biomass, while the ISE of weed-free yield was 1.05 and 0.77, respectively. On a single-year basis, the ISE of crop vigor at 4 WAS was 1.14 for crop yield under weed competition and 1.38 for weed biomass, while the ISE of weed-free yield was 1.40 and 1.11, respectively (Table 1). Because of the high ISEs of weed-free yield and early vigor resulting from their high H and genetic correlations with the two target weedy traits, they are identified as the most promising traits that can be used under weed-free conditions as indirect selection criteria for yield and weed biomass under weedy conditions. The relatively higher ISEs on a single year basis than those on a multi-year basis were probably from their higher predicted broad-sense heritabilities, which were biased upward because of the confounding of genotype and genotype × year variances. However, the degree of bias is similar across traits, and therefore is unlikely to affect the comparisons among traits for inclusion in indirect selection indices.

Indirect Selection by Index and Independent Culling Levels

Through stepwise regression analysis, the crop grain yield under weed competition ($Yield_{Weedy}$) and weed biomass (Weed) were predicted using crop grain yield under weed-free conditions ($Yield_{Weed\text{-}free}$) and crop seedling vigor rated at 4 WAS under weed-free conditions ($Vigor_{Weed\text{-}free}$) as independent predictor variables (Eqs. (1) and (2)). The two weed-free traits, grain yield and vigor, together explained 89% of variation in yield and 48% in weed biomass in weedy environments. Therefore, indirect selection under weed-free conditions for both yielding ability and WSA can be effective.

$$Yield_{Weedy} = 0.01 + 0.38\ Yield_{Weed\text{-}free} + 0.06\ Vigor_{Weed\text{-}free} \qquad R^2=0.89 \qquad (1)$$

$$Weed = 244.84 - 5.92\ Yield_{Weed\text{-}free} - 10.47\ Vigor_{Weed\text{-}free} \qquad R^2=0.48 \qquad (2)$$

$$\text{Yield}_{\text{Weedy}} = -0.82 + 0.44\ \text{Yield}_{\text{Weed-free}} + 0.03\ \text{Height}_{\text{Weed-free}} \qquad R^2=0.89 \qquad (3)$$

$$\text{Weed} = 417.16 - 13.34\ \text{Yield}_{\text{Weed-free}} - 5.31\ \text{Height}_{\text{Weed-free}} \qquad R^2=0.55 \qquad (4)$$

Regression analysis with inclusion of just a single independent variable showed that weed-free yield is more important than weed-free crop vigor in predicting yield under competition, while weed-free crop vigor is more important than weed-free yield in predicting weed biomass (not shown). Thus, selecting on both yield and vigor under weed-free conditions is necessary to simultaneously improve yielding ability under weed competition and WSA. It was also observed that weed-free yield and early vigor were positively correlated, implying no trade-off between the two selection criteria. The two criteria may serve in one selection index with the regression coefficients as their weights. Use of such a selection index, however, would require the harvest of all the lines, sometimes several thousands, in the pedigree nurseries, requiring a great deal of labor. Selection using independent culling levels (Bernardo, 2002) may be more practical. Within one season, the first selection is conducted on seedling vigor at 4 WAS, followed by the second selection based on grain yield. In this case, harvest of entries that exhibit high vigor only is needed, thus saving labor and allowing breeders to manage larger breeding populations. In addition, early plant height (at 4 WAS) was found to be as good as early crop vigor in predicting crop yield under weed competition and weed biomass (Eqs. (3) and (4)). Therefore, vigor evaluation may be replaced by height measurement for the early season selection in case poor crop establishment does not allow a proper vigor rating. But measuring plant height is more laborious and slower than visually rating seedling vigor.

The selection protocols for WC were developed based on a large population of diverse genotypes, with clear differences in traits among cultivars. Its applicability in a narrower population of progenies from crosses among parents with similar growth characters is not known and requires further study. However, since weed-competitive aerobic rice breeding mainly aims at improving drought tolerance, yield and WSA, often the crosses made will be between genotypes which are drought-tolerant and weed-suppressive (upland or aerobic rice), and genotypes which are high yielding (lowland rice). The genetic variability within the segregating populations derived from such crosses is expected to be large enough to permit effective selection. The application of this protocol in IRRI's current aerobic rice breeding programs has proven this. A study in wheat (Mokhtari et al., 2002) also showed that variability in plant growth and yield within an F_3 population is large enough for selecting promising lines with respect to WC. Jannink et al. (2000) reported that selection on early height can be effective for improving WSA in soybean, but he argued that gains for yield from selection may be difficult while improving WSA because of a negative correlation between early height and yield. However, this is unlikely to happen in aerobic rice because of the positive association between the two target traits, yielding ability and WSA, and between the two selection criteria and the two target traits, as discussed earlier.

RELATIONSHIP BETWEEN PLANT TYPE AND WEED-SUPPRESSIVE ABILITY

One interesting finding in our research is that the droopy plant types tend to be less weed-suppressive than the erect ones. This conclusion is based on the following results:

- In the population of 40 diverse genotypes, visual ratings of plant erectness were negatively correlated with weed biomass, but positively correlated with crop vigor, vegetative crop biomass and yield, which were also negatively correlated with weed biomass.
- *Indica* and *aus* germplasm groups were both erect and more weed-suppressive than the relatively droopy *japonica* germplasm groups.
- Within *japonica* germplasm, plant erectness was also negatively but not significantly correlated with weed biomass (not shown).
- In the experiment with three genotypes, the erect genotype Apo was more weed-suppressive than the other two relatively droopy genotypes IRAT216 and IR60080-46A.

The droopy plant type was hypothesized to provide larger GC, thus enabling it to suppress weeds more than the erect plant type, but this research showed that there was no close relationship between GC and plant erectness, and that the erect *indica* and *aus* germplasm groups even had a somewhat larger GC than the droopy *japonica* groups. These results indicate that the above-mentioned hypothesis may be not true. Under field conditions, with constant planting density and row spacing, GC of a cultivar is determined by its growth rate, LAI, tillering, tiller angle, plant height, and leaf erectness. Therefore, the contribution of droopy leaves to GC may be very limited. Audebert et al. (1999) reported that the greater GC of *O. glaberrima* cultivars relative to *O. sativa japonica* cultivars is achieved by rapid increase in the number of leaves through high tillering, but not through characteristics of individual leaves. Dingkuhn et al. (1999) observed the same light extinction coefficients at 4.5 WAS for *O. glaberrima* and *O. sativa* cultivars with contrasting plant types. These studies indicate that effect of plant erectness on GC is small. However, the observed negative effect of droopy plant type on WSA may not result from the droopy plant type *per se*. The weak negative association between erectness and weed biomass within *japonica* germplasm (not shown) suggests that droopy plant type *per se* is a trait that is unrelated to WSA. The negative association between droopy plant type and WSA observed in the diverse 40-cultivar population and among the germplasm groups may result from the fact that all the cultivars belonging to *indica* and *aus* germplasm groups used in this research are erect and have fast early growth. It may have been a chance occurrence that the *japonica* materials included in this study did not have fast early growth, in which case, droopy plant type *per se* may not be detrimental to WSA. However, as discussed earlier, fast early growth is the most important attribute of a strongly weed-suppressive genotype. This viewpoint is supported by Gibson and Fischer (2001), who found that shade alone does not prevent the weed growth due to morphological plasticity and changes in dry matter allocation of weeds expressed under shading stress, and who speculated that fast nutrient deprivation of weeds by rice is more crucial to weed suppression. Similarly, Johnson et al. (1998) linked the advantage in WSA of *glaberrima* over *japonica* cultivars to early tiller production and early biomass accumulation. In contrast, however, studies with rice (Dingkuhn et al., 1999) and wheat (Lemerle et al., 1996) suggest that droopier cultivars are more weed-suppressive.

EFFECT OF INTERACTION OF PLANT DENSITY AND GENOTYPE ON WEED-SUPPRESSIVE ABILITY

The study with three genotypes differing in WSA grown at the seeding rates of 100, 300 and 500 viable seeds m^{-2} revealed that there was no significant interaction between plant density and genotype for yield and WSA; in other words, the effects of genotype and plant density were additive. Both weed-suppressive genotypes and increased seeding rates decreased both weed growth and yield loss. Yield under competition and weed suppression of a genotype increased with increased seeding rate from 100 to 300 viable seeds m^{-2}, but did not do so with further increase in seeding rate over 300 viable seeds m^{-2}, indicating that a rate of 300 viable seeds m^{-2} is appropriate in terms of crop yield and weed suppression under moderate weed infestation levels regardless of strong or weak WSA of a genotype. It was also found that a strongly weed-suppressive genotype at a lower seeding rate can be equivalent in weed suppression to a weakly weed-suppressive genotype at a higher seeding rate, indicating that weakly weed-suppressive genotypes need to be grown at a relatively high seeding rate to achieve satisfactory weed suppression. The strongly weed-suppressive genotype was found in this experiment, as in Experiment 1, to have fast early vegetative growth, expressed as rapid tillering, vegetative biomass accumulation, and LAI development, thus allowing early canopy closure. In summary, a weed-suppressive genotype combined with an appropriate seeding rate (300 viable seeds m^{-2}) could effectively suppress the growth of naturally-germinating weeds. This may be a strategy that minimizes weeding operations in aerobic rice fields to one time per season, freeing labour from onerous weeding practices and reducing dependency on herbicides.

Farmers often plant extremely large amounts of seed to facilitate weed management. However, an excessive seeding rate which results in severe intra-specific competition may reduce grain setting of a crop, causes lodging (Bond et al., 2005), more severe rat damage (Castin and Moody, 1989) and increased insect and disease damage (Tan et al., 2000), and consequently reduce final yield. Use of weed-suppressive cultivars with rapid early growth can permit lower seeding rates to be used without loss of WSA, avoiding these negative consequences of high seeding rates.

CONCLUSION

This research shows that (1) weed-suppressive genotypes are effective in aerobic rice weed management, especially so when grown at an appropriate seeding rate (300 viable seeds m^{-2}); (2) there is a large genetic variability in WSA in aerobic rice, permitting genetic improvement for WSA; (3) positive association between yielding ability under weed competition and WSA allows simultaneous improvement of the two traits; and (4) selection for WC can be conducted indirectly on yield and early crop vigor under weed-free conditions.

According to these findings, WC should be included among the main breeding goals together with drought tolerance and yield potential in aerobic rice.

With the challenge of feeding more people on a rice-based diet in the decades ahead under increasing water scarcity, a breakthrough in yield and WSA by targeted aerobic rice breeding would provide benefits in water saving while increasing rice production. This goal is achievable according to our study. With the expected high-yielding weed-suppressive aerobic rice genotypes, irrigated lowland rice may be partly replaced by direct-seeded aerobic rice, especially in water scarce areas in Asia. Traditional low-yielding upland rice may also be replaced with high-yielding aerobic rice in areas where rainfall is uniform and frequent during a growing season. Our results indicate that these changes may occur without greatly increased use of herbicides. Rice production may, therefore, go on in a more resource-saving, environment-friendly and development-sustainable way.

REFERENCES

Anonymus (1993). Rice Almanac. International Rice Research Institute, Los Baños, Philippines.

Atlin, G.N., Cooper, M., Bjørnstad, Å. (2001). A comparison of formal and participatory breeding approaches using selection theory. *Euphytica 122*, 463-475.

Audebert, A., Dingkuhn, M., Jones, M.P., Johnson, D.E. (1999). Physiological mechanisms for vegetative vigour of interspecific upland rice: Implications for weed competitiveness. In: Proc. Intern. Symp.: World Food Security, Kyoto, 300-301.

Bastiaans, L., Kropff, M.J., Kempuchetty, N., Rajan, A., Migo, T.R. (1997). Can simulation models help design rice cultivars that are more competitive against weeds? *Field Crops Res. 51*, 101-111.

Bernardo, R. (2002). Selection for multiple traits. In: Bernardo, R. (Ed.), *Breeding for quantitative traits in plants*. Stemma Press, Woodbury, Minnesota, 261-274.

Bond, J.A., Walker, T.W., Bollich, P.K., Koger, C.H., Gerard, P. (2005). Seeding rates for stale seedbed rice production in the midsouthern United States. *Agron. J. 97*, 1560-1563.

Castin, E.M., Moody, K. (1989). Effect of different seeding rates, moisture regimes, and weed control treatments on weed growth and yield of wet-seeded rice. In: Proceedings 12th Asian-Pacific Weed Sci. Soc. Conf., Seoul, Korea, 337-343.

Caton, B.P., Cope, A.E., Mortimer, M. (2003). Growth traits of diverse rice cultivars under severe competition: Implications for screening for competitiveness. *Field Crops Res. 83*, 157-172.

Cooper, M., Delacy, I.H., and Basford, K.E. (1996). Relationships among analytical methods used to analyze genotypic adaptation in multi-environment trials. In: M. Cooper and G.L. Hammer (Eds.), *Plant adaptation and crop improvement*. CAB Int., Wallingford, UK, 193-224.

De Datta, S.K., Llagas, M.A. (1984). Weed problems and weed control in upland rice in tropical Asia. In: An overview of upland rice research. IRRI, Los Baños, Laguna, Philippines, 321-342.

Dingkuhn, M., Johnson, D.E., Sow, A., Audebert, A.Y. (1999). Relationships between upland rice canopy characteristics and weed competitiveness. *Field Crops Res. 61*, 79-95.

Falconer, D.S. (1989). Introduction to quantitative genetics. 3rd ed. Longman Scientific and Technical, Essex, England.

Fischer, A.J., Chatel, M., Ramirez, H., Lozano, J., Guimaraes, E. (1995). Components of early competition between upland rice (*Oryza sativa* L.) and *Brachiaria brizantha* (*Hochst. Ex A.Rich*) Stapf. *Int. J. Pest Manag. 41*, 100-103.

Fischer, A.J., Ramirez, H.V., Lozano, J. (1997). Suppression of junglerice (*Echinochloa colona* (L.) Link) by irrigated rice cultivars in Latin America. *Agron. J. 89*, 516-521.

Fischer, A.J., Ramirez, H., Gibson, K.D., Pinheiro, B.D.S. (2001). Competitiveness of semidwarf upland rice cultivars against palisadegrass (*Brachiaria brizantha*) and signalgrass (*B. decumbens*). *Agron. J. 93*, 967-973.

Fofana, B., Rauber R. (2000). Weed suppression ability of upland rice under low-input conditions in West Africa. *Weed Res. 40*, 271-280.

Garris, A.J., Tai, T.H., Coburn, J., Kresovich, S., McCouch, S. (2005). Genetic structure and diversity in *Oryza sativa* L. *Genetics 169*, 1631-1638.

Garrity, D.P., Movillon, M., Moody, K. (1992). Differential weed suppression ability in upland rice cultivars. *Agron. J.* 84, 586-591.

Gibson, K.D., Fischer, A.J. (2001). Relative growth and photosynthetic response of water-seeded rice and *Echinochloa oryzoides* (Ard.) Fritsch to shade. *Intern. J. Pest Manag. 47*, 305-309.

Gibson, K.D., Fischer, A.J. (2004). Competitiveness of rice cultivars as a tool for crop-based weed management. In: Inderjit (Ed.), *Weed biology and management*. Kluwer Academic Publishers, The Netherlands, 517-537.

Gibson, K.D., Hill, J.E., Foin, T.C., Caton, B.P., Fischer, A.J. (2001). Water-seeded rice cultivars differ in ability to interfere with watergrass. *Agron. J. 93*, 326-332.

Gibson, K.D., Fischer, A.J., Foin, T.C., Hill, J.E. (2003). Crop traits related to weed suppression in water-seeded rice (*Oryza sativa* L.). *Weed Sci. 51*, 87-93.

Glazsmann, J.C. (1987). Isozymes and classification of Asian rice varieties. *Theor. Appl. Genet. 74,* 21-30.

Goldberg, D.E., Landa, K. (1991). Competitive effect and response: Hierarchies and correlated traits in the early stage of competition. *J. Ecology 79*, 1013-1030.

Haefele, S.M., Johnson, D.E., Bodj, D.M., Wopereis, M.C.S., Miezan, K.M. (2004). Field screening of diverse rice genotypes for weed competitiveness in irrigated lowland ecosystems. *Field Crops Res. 88*, 39-56.

Heap, I. (2006). The international survey of herbicide resistant weeds. Available at www.weedscience.com

Jannink, J.-L., Orf, J.H., Jordan, N.R., Shaw, R.G. (2000). Index selection for weed-suppressive ability in soybean. *Crop Sci. 40*, 1087-1094.

Jennings, P.R., Jesus Jr, J.D. (1968). Studies on competition in rice. I. Competition in mixtures of varieties. *Evolution 22*, 119-124.

Jennings, P.R., Herrera, R.M. (1968). Studies on competition in rice. II. Competition in segregating populations. *Evolution 22*, 332-336.

Jennings, P.R., Aquino, R.C. (1968). Studies on competition in rice. III. The mechanism of competition among phenotypes. *Evolution 22*, 529-542.

Johnson, D.E. (1996). Weed management in small holder rice production in the tropics. Jan. 24, 2006. Available at http://ipmworld.umn.edu/chapters/johnson.htm.

Johnson, D.E., Dingkuhn, M., Jones, M.P., Mahamane, M.C. (1998). The influence of rice type on the effect of weed competition on *Oryza sativa* and *Oryza glaberrima*. *Weed Res. 38*, 207-216.

Jordan, N. (1993). Prospects for weed control through interference. Ecol. Appl. 3, 84-91.

Kawano, K., Gonzalez, H., Lucena, M. (1974). Intraspecific competition, competition with weeds, and spacing response in rice. *Crop Sci. 14*, 841-845.

Lemerle, D., Verbeek, B., Cousens, R.D., Coombes, N.E. (1996). The potential for selecting wheat varieties strongly competitive against weeds. *Weed Res. 36*, 505-513.

Lemerle, D., Verbeek, B., Orchard, B. (2001[a]). Ranking the ability of wheat varieties to compete with *Lolium rigidum*. *Weed Res. 41*, 197-209.

Lemerle, D., Gill, G.S., Murphy, C.E., Walker, S.R., Cousens, R.D., Mokhtari, S., Peltzer,S.J., Coleman, R., Luchett, D.J. (2001[b]). Genetic improvement and agronomy for enhanced wheat competitiveness with weeds. *Aust. J. Agric. Res. 52*, 527-548.

Lotz, L.A.P., Wallinga, J., Kropff, M.J. (1995). Crop-weed interactions: Quantification and prediction. In: Glen, D.M., Greaves, M.P., Anderson, H.M. (Eds.), *Ecology and Integrated Farming Systems*. John Wiley & Sons Ltd., 31-37.

Mokhtari, S., Galwey, N.W., Cousens, R.D., Thurling, N. (2002). The genetic basis of variation among wheat F3 lines in tolerance to competition by ryegrass (*Lolium rigidum*). *Euphytica 124*, 355-364.

Ni, H., Moody, K., Robles, R.P., Paller, E.C., Lales, J.S. (2000). *Oryza sativa* plant traits conferring competitive ability against weeds. *Weed Sci. 48*, 200-204.

Ni, H., Moody, K., Robles, R.P. (2004). Analysis of competition between wet-seeded rice and barnyardgrass (*Echinochloa crus-galli*) using a response-surface model. *Weed Sci. 52*, 142-146.

Nyquist, W.E. (1991). Estimation of heritability and prediction of selection response in plant populations. *Crit. Rev. Plant Sci. 10*, 235-322.

Oerke, E.C., Dehne, H.W. (2004). Safeguarding production-losses in major crops and the role of crop protection. *Crop Prod. 23*, 275-285.

Oka, H. (1960). Variation in competitive ability among rice varieties. *Jap. J. Breeding 10*, 61-68.

Phuong, L.T., Denich, M., Vlek, P.L.G., Balasubramanian, V. (2005). Suppressing weeds in direct-seeded lowland rice: Effects of methods and rates of seeding. *J. Agron. Crop Sci. 191*, 185-194.

Roder, W., Keobulapha, B. (1997). Weeds in slash-and-burn rice fields in northern Laos. *Weed Res. 37*, 111-119.

SAS Institute Inc. (1999-2001). SAS System Version 8. SAS Institute Inc., Cary, NC.

Singh, S., Ladha, J.K., Gupta, R.K., Bhushan, L., Rao, A.N. (2008). Weed management in aerobic rice systems under varying establishment methods. *Crop Protection 27*, 660-671.

Tan, P.S., Khuong, T.Q., Hoai, N.T. (2000). Low cost technologies for rice production in the Mekong delta. Paper presented at the National Workshop on 21-23 Sept. 2000 in Ho Chi Minh City, Vietnam, 14 p.

Tuong, T.P., Bouman, B.A.M. (2003). Rice production in water-scarce environments. In: Kijne, J.W., Barker, R., Molden, D. (Eds.), *Water productivity in agriculture: Limits and opportunities for improvement*. CABI Publishing, 53-67.

Wall, P.C. (1983). The role of plant breeding in weed management in the advancing countries. In: Improving weed management. Proc. FAO/IWSS Expert Consultation on Improving Weed Management in Developing Countries. Rome, 6-10 Sept. 1982, FAO, Rome, 40-46.

Wang, H.Q., Bouman, B.A.M., Zhao, D.L., Wang, C., Moya, P.F. (2002). Aerobic rice in northern China: Opportunities and challenges. In: Bouman, B.A.M., Hengsdijk, H., Hardy, B., Bindraban, P.S., Tuong, T.P., Ladha, J.K. (Eds.), Water-wise rice production. Proc. Intern. Workshop on Water-wise Rice Production, 8-11 April 2002, Los Baños, Philippines. International Rice Research Institute, 143-154.

Zhao, D.L., Bastiaans, L., G.N. Atlin, G.N., & Spiertz, J.H.J. (2007). Interaction of genotype x management on vegetative growth and weed suppression of aerobic rice. *Field Crops Res. 100*, 327-340.

Zhao, D.L., Atlin, G.N., Bastiaans, L., & Spiertz, J.H.J. (2006[a]). Cultivar weed-competitiveness in aerobic rice: heritability, correlated traits, and the potential for indirect selection. *Crop Sci. 46*, 372-380.

Zhao, D.L., Atlin, G.N., Bastiaans, L., & Spiertz, J.H.J. (2006[b]). Developing selection protocols for weed competitiveness in aerobic rice. *Field Crop Res. 97*, 272-285.

Zhao, D.L., Atlin, G.N., Bastiaans, L., & Spiertz, J.H.J. (2006[c]). Comparing rice germplasm groups for growth, grain yield, and weed-suppressive ability under aerobic soil conditions. *Weed Res. 46*, 444-452

Zhao, D.L. (2006[d]). *Weed competitiveness and yielding ability of aerobic rice genotypes*. PhD thesis Wageningen University, ISBN: 90-8504-410-3.

Zimdahl, R.L. (2004). *Weed-crop competition: A review.* Second edition. Blackwell Publishing, 2121 State Avenue, Ames, IA50014.

Note: This paper is a summary of a PhD research. Funding for this research was provided by the International Rice Research Institute (IRRI) in the Philippines and a grant for a PhD-fellowship from Wageningen University, The Netherlands.

In: Weeds: Management, Economic Impacts and Biology
Editor: Rudolph V. Kingely
ISBN: 978-1-60741-010-2

Chapter 3

DEVELOPING WEED CONTROL METHODS FOR CAMELINA: A NEWLY INTRODUCED, HIGH-VALUE OIL SEED CROP

Susanne Heiska[1]
Department of Applied Biology,
P.O. Box 27, 00014 University of Helsinki, Finland.

ABSTRACT

Camelina, or gold-of-pleasure (*Camelina sativa* L. Crantz), is an oil-seed crop with a cultivation history dating back to the Neolithic era. Later, the crop was superseded by other oil seed crops such as rape or turnip rape, but during the last decade camelina has awoken new interest; recent studies show the seed oil having extraordinary good fatty-acid composition and thus being a high-value ingredient in functional food supplies. In Finland, the cultivation of camelina started in the late 1990s and the area under the crop has been growing since then; in 2006 it was cultivated on more than 5,000 ha. The weeds, however, have been found to be challenging for cultivation. As a recently introduced crop cultivated in relatively small area, there are no herbicides developed for camelina. The weeds have been controlled using herbicides such as trifluralin or metazachlor, commonly used for rape. Metazachlor is, however, known to easily injure crops belonging to the Brassicaceae family, and the use of trifluralin has been debated because of its persistence in the soil. There are also growing demand and markets for functional food supplies cultivated organically.

To develop new weed control methods for camelina, different combinations of the timing of weed control by spring-tine harrow, sowing rate and soil type were tested in a two-year field trial conducted in Helsinki, Finland. The study included two cultivars, one sown in the spring and the other in the fall. The effects of different mechanical weed control methods on the yield and above-ground biomass of the weeds were measured at-harvest, and the number of the weeds was recorded several times during the growing seasons. The effects of different mechanical weed control methods were compared with herbicide treatment with trifluralin and hand-weeded and unweeded control treatments.

---

[1] Email: susanne.heiska@gmail.com

The results show that the yield of camelina was more dependent on the year and the soil type than on the weed control method. In comparison with unweeded control, 0–83% of the weeds were controlled by mechanical weed control methods depending on the soil type, sowing density and the timing of the weed control practices. The efficacy of most of the mechanical weed control methods was comparable to trifluralin treatment. However, the yield in the hand-weeded control plots was higher compared with the mechanically or chemically controlled plots, indicating that the yield could be increased by weed control methods developed for camelina.

INTRODUCTON

Botanical Description

Gold-of-pleasure, also known as false-flax or linseed dodder (*Camelina sativa* L. Crantz), has its origin in the Mediterranean and Central Asia (Putnam et al. 1993). The golden coloured and oval shaped seed of camelina is small, with 1000-seed weight typically ranging between 0.9 g and 1.6 g (Gugel and Falk 2006; Vollman et al. 2007). During the early vegetative phases, the first leaves of camelina form a rosette that is considered beneficial for competing with the weeds. In field conditions, the flowering plant usually reaches a height of 30 to 90 cm, but in the native growing habitat the plants typically grow lower (Plessers et al. 1962; Putnam 1993). Camelina is self-pollinated and has small, pale yellow flowers with four petals typical of the Brassicaceae family (Fig. 1). The flowering does not require high temperatures (Plessers et al. 1962), which makes the cultivation and seed formation possible even in the boreal conditions. After the flowering, the seeds are formed within a pear-shaped pod. The seeds contain approximately 40% oil with a high content of linolenic acid (Schuster and Friedt 1995; Zubr 2003). During ripening, the colour of the pod turns from light green to golden and finally greyish brown. The growing season ranges between 75 and 110 days (Rüther 1957; Plessers et al. 1962).

Figure 1. The flowering camelina.

The History of Cultivation and Use of Camelina

The first records of camelina cultivation date back to the Neolithic era. The high content of oil in the seeds was known ca. 4,000 years ago when the plant was cultivated in Eastern and Central Europe and Scandinavia (Büchsenschütz-Nothdurft et al. 1992; Dönmez and Belli 2007). During the Roman era, the cultivation was extensive in Europe, but little by little camelina was superseded by other, higher-yield oil-seed crops (Putnam et al. 1993). During the 20th century, camelina was cultivated to a certain extent in northern France and in the area of the former Soviet Union (Plessers et al. 1962). During prehistoric times, camelina seeds were ground up and used to give a pleasant, nutty taste to bread and porridges. The oil pressed from the seeds was later noticed to smoke less than oil derived from the seeds of other Brassica and thus, during Roman times the oil was commonly used in oil lamps (Fågelfors 1984). During the 19th century, camelina was known as an herbal remedy; in folk medicine the leaves, seeds and oil were used in treating, e.g., eye inflammation, burns and stomach ulcers (Rode 2002).

Cultivation Today

Recently, camelina has awoken new interest because of its high potential as a high-quality comestible oil and energy crop. The cold-pressed camelina oil is an interesting source of relatively stable omega 3 fatty acids with valuable properties that lower cholesterol in human nutrition (Karvonen et al. 2002; Zubr 1992). The oil also has potential as an ingredient in paints and cosmetics with low environmental load and as a source of biofuel (Bonjean and Le Geoffic 1999; Bernardo et al. 2003; Fröhlich and Rice 2003). In Finland, the oil is used as an ingredient in functional food supplies such as margarine, cooking oil or unripened cheese. The suitability of oilseed cake, with a relatively high content of seed oil left over, for animal nutrition has been well tested (e.g., Moloney et al. 1998; Peiretti et al. 2007). The use of camelina for animal nutrition is prohibited by an EU directive. However, the evidence from recent studies repeatedly shows that a small portion of 5% camelina cake can be added to poultry diet without any implications for animal health or quantity of feeding (Frame et al. 2007; Weissmann et al. 2007). This amount, however, is proven to enhance the meat quality and produce added value to both camelina cultivation and meat and egg production (Rokka et al. 2002; Ryhänen et al. 2007). Under the existing EU legislation, by far the most profitable use of the oil seed cake has probably been as soil enrichment (Prankl and Rathbauer 1997).

Because of the multifaceted applications of camelina, the potential of its cultivation has been tested during the last decade in numerous field studies conducted in Central Europe, Fenno-Scandia, Canada and Australia (e.g., Müller et al. 1999; Francis and Campbell 2003; Zubr 2003; Gugel and Falck 2006). In Finland, the commercial cultivation of camelina started in 1999 with a few hundred hectares and since then the area under cultivation has increased—in 2006, camelina was cultivated on more than 5,000 ha in Finland. Typically, the yield of camelina ranges between 1.2 and 2.0 Mg ha^{-1}. The hectare yield of camelina is lower than that of other oil seed crops cultivated in Finland (turnip rape, rape and flax), but the relatively high prices have motivated farmers to try out a new crop. Introduction of camelina has been quite easy, as the crop can be cultivated with the machinery typically used on Finnish farms. Despite the small size of the seed, the establishment of camelina has been shown to be reliable when sown at a depth of 1–2 cm. In cultivation, camelina has shown extraordinary resistance to several pathogens infecting the Brassicaceae family, such as *Alternaria spp.* (e.g., Sigareva and Earle 1999; Westman et al. 1999; Vollmann et al. 2001). Camelina has also shown a high resistance to flea beetles (*Phyllotreta cruciferae*) that may cause serious injury in rape (Pachagounder et al. 1998; Henderson et al. 2004). Farmers have also described the harvesting being easier than that of rape or turnip rape.

The biggest problem in camelina cultivation has been the weeds—there are no weed control methods developed for camelina. In uncontrolled vegetation, weed competition may reduce the camelina yield. The weeds may also slow down the growth of camelina and cause uneven ripening of the crop. The chlorophyll content of the unevenly ripened seeds may vary and lower the yield quality. The weed seeds harvested with the camelina yield may contain undesired constituents and should therefore be removed from the yield. The sorting of the seeds is, however, difficult and time consuming because of the small size of camelina seed. Thus, developing an efficient weed control method for camelina would be essential for the future possibilities in production.

Weed Management

Herbicides such as metazachlor, trifluralin or napropamide have been commonly used for the control of weeds on camelina. However, the use of metazachlor cannot be recommended, because it slows down the crop growth. Trifluralin, instead, has been widely used for camelina, and it has been controlling effectively a wide range of the weed species commonly found in camelina. On the other hand, trifluralin is known to degrade slowly in the soil, and currently its marketing and use as herbicide in Finland has been prohibited since the 2007. In practice, the effect of napropamide on some weeds with the seed size similar to camelina, such as fat hen (*Chenopodium album*) or field pennycress (*Thlaspi arvense*) has been moderate or weak and the herbicide is therefore not considered adequate as the only weed control method for camelina. As the selection of the herbicides suitable for camelina is very limited, a solution to the weed problem was searched amongst the mechanical weed control methods.

The mechanism of mechanical weed control can be based on cutting or slicing the weed at such stage of growth in which the storage below the cutting point is not sufficient for the re-growth and efficient expansion of the leaf area. The mechanism can also be based on pulling or uprooting the weeds or burying them by using different machines developed for this purpose (Lötjönen 1995; Fogelberg and Dock Gustavsson 1998). Many of the mechanical weed control methods have been developed, especially for the inter-row weed control, but some of the methods also control the weeds in intra-rows (Kouwenhoven 1997). Selectivity of the mechanical weed control can be defined as the ratio between weed control and crop injury (Rasmussen 1992). In general, mechanical weed control is most selective when the crops differ from weeds in their growth habit, time of emergence or the time of maturity (Rasmussen 1990, 1991). Weeds are, however, superior in adapting to cultivation practices and sometimes injury to the crop is a prerequisite for the mechanical weed control. The competition ability of the crop can, however, be increased by increasing the sowing density (Lawson and Topham 1985; Ghersa and Roush 1993). Both the mechanism of weed control and the selectivity of control method are important factors that should be considered when choosing the mechanical weed control method matching the cultural practices and growth habit of the crop.

Spring-tine harrowing is a relatively cheap method controlling the weeds both in inter-rows and intra-rows (Baumann 1992). Weed harrowing can be used for a wide range of crop plants, including e.g. cereals, sugar beet, potato, pea or perennial grasses (Rasmussen 1992; Kouwenhoven 1997; Hansen et al. 2007). Weed harrowing also controls effectively annual weed species; the controlling effect may range between 20% and 80% (Rasmussen 1992; Pullen and Cowell 1997). The mechanism of weed control by the spring-tine harrow is mainly based on the soil movement caused by the tine (Kurstjens and Kropff 2001). Vibration of the flexible tines can effectively remove soil from the uprooted weeds and thus prevent the re-growth after the harrowing (Pullen and Cowell 1997). The tine can also cut the weeds in direct contact between the tine and weeds (Duerinckx et al. 2005).

The intensity of spring-tine harrowing depends on various controllable and uncontrollable factors, such as harrow adjustment, the timing of harrowing, the growth stage of the crop plant and weeds, the driving speed and direction, the number of passes, soil type and water relations and weather conditions (e.g. Rydberg 1993; Kurstjens et al. 1999; Duerinckx et al. 2005). The selectivity of the harrowing is also determined by these same factors; increased intensity often declines the selectivity (e.g., Rasmussen 1990; Kurstjens and Kropff 2001; Duerinckx et al. 2005). Thick and moderately rigid tines control the weeds effectively, but harrows with thin and flexible tines may be very selective. The flexibility of the tine, however, complicates adjustment of the harrowing depth. Selective harrowing, on the other hand, requires minimising the tine upward movement (Duerinckx et al. 2005). The position of tines is normally upright but can be adjusted to slightly leading or trailing. The tine movement and its effect on the soil particles can be adjusted by changing the position but also by increasing driving speed (Rydberg 1993; Kurstjens et al. 1999; Duerinckx et al. 2005). The increased driving speed of harrow with tines in the leading position reduces the selectivity more rapidly than with tines in the upright position (Kouwenhoven 1997).

A 'blind harrowing' before the crop emergence can selectively control the weeds germinating in the soil profile above the crop (Rasmussen and Rasmussen 1999). Harrowing controls effectively annual weeds at cotyledon stage, but at this time the crop can be easily injured (Rydberg 1993; Rasmussen and Svenningsen 1995). Well established crops can, however, pull the flexible tines aside of the row and thus, the selectivity of harrowing performed along the crop rows increases as the crops grow (Vanhala 1992). One pass is seldom sufficient as the only method for weed control, but increasing the number of passes may, increase the risk of crop injury (Rydberg 1991). In rainy growing seasons, the timing of the harrowing may be very challenging. In wet conditions, the selectivity of harrowing is low and the uprooting of weeds is poor (Koch 1964; Rasmussen 1990; Wilson et al. 1993 but see also Kurstjens et al. 1999). In all, the successful weed harrowing requires a careful adjustment of the controllable factors to the weather conditions and soil type.

Objectives of the Study

This study aimed to evaluate the possibilities of spring-tine harrowing controlling the weeds on annual and winter-annual camelina. The growing habit of camelina is known to differ from the crops typically cultivated in Finland. Emergence of the crop is rapid; normally the cotyledons emerge in a few days after the sowing. The first true leaves of camelina form a rosette which usually is an adaptive trait indicating high competitive ability. According to the farmers experience in Finnish conditions, the growth of camelina, however, slows down at the rosette stage giving an advantage to the weeds with the rapid emergence and establishment. In practice, the lately emerging weeds such as fat hen and scentless may weed (*Tripleurospermum inodorum*) have been the most distracting in camelina cultivations. A solution to the weed problems was sought in two-year field trials comprising different combinations of sowing density and mechanical weed control methods with spring-tine harrow.

The objectives of this study were to

i. compare the effects of the timing and the number of passes with spring-tine harrow on the weed control and the selectivity of harrowing and the yield of camelina
ii. compare the success of mechanical weed control with chemical control
iii. evaluate the weed control-by-year-by-soil type interaction with a purpose of fitting the controllable factors to the uncontrollable factors determining the intensity and selectivity of spring-tine harrowing.

The assumption of this study was that a harrowing made at early stage (1-2 leaves) of camelina would effectively control the annual weeds commonly competing with camelina. The control effect of the harrowing replicated at the rosette stage (4-6 leaves) of camelina was also assumed to effectively control the weeds at cotyledon stage. The selectivity of the harrowing was assumed being poor prior to the rosette stage of camelina. The increased sowing density was, however, assumed to compensate the injury caused by the harrowing.

Material and Methods

Field experiments were conducted during 1998-1999 in Viikki Experimental Farm, Helsinki Finland, using two camelina cultivars delivered by Prof. Zubr, The Royal Veterinary and Agricultural University, Denmark. One of the cultivars was annual and the other was winter-annual. The annual cultivar was cultivated for the growing seasons 1998 and 1999 in two soil-types. The winter annual cultivar was cultivated for one growing season, 1998–1999, in one soil-type.

Experiments with Annual Cultivar

Prior to the sowing of annual cultivar, the soil was ploughed and harrowed with the purpose of assuring an even germination of the small seeds. The study area was fertilised with 100 kg ha^{-1} N. Camelina seeds were sown in rows with the spacing of 12.5 cm using two sowing densities, 8 and 11 kg ha^{-1}. In 1998 and 1999, the annual cultivar was sown on clay on May 15 and May 12, respectively, and on very fine sand on May 20 and May 14, respectively. The seeds were sown in plots with the size of 1.25 X 8.00 m. The experiments on clay in 1998 and 1999 were located in the same brake, but the experiment on very fine sand in 1998 was located in another brake than in 1999.

Weed Control Methods

The weed control methods included different weed harrowing treatments, a chemical control, a hand-weeded control and an unweeded control. The harrowing treatments included three (in 1998) or five (in 1999) different combinations of timing and number of passes by harrow. All the harrowing treatments were made by a spring-tine harrow (manufactured by Rakokoneet, Mankila) with the working width of 1.5 m (Fig. 2a). The tines of the harrow were mounted in four bars with the distance of 10 cm between each of the tines in the same bar. The position of the tines in the bars was overlapping, so that the working width of one tine was 3 cm. The harrowing was done along the camelina rows and the tines were adjusted on the slightly trailing position during the harrowing (Fig. 2b). The driving speed was 4-6 km h^{-1}. The weed control methods were replicated in the two different sowing densities in clay and very fine sand in 1998 and 1999.

Figure 2. The spring-tine harrow manufactured by Rakokoneet, Mankila was used for the mechanical control of weeds on camelina (a). The position of the flexible tines was adjusted to be slightly trailing during the harrowing treatment (b).

The timing of harrowing treatments was adjusted based on the growth stage of camelina. In half of the plots planned to be harrowed, the first pass by harrow was performed when the first leaves of camelina appeared between the cotyledons (early harrowing). In the other half of the plots planned to be harrowed, the first pass was done when the plants had a rosette formed of 4–6 leaves (late harrowing). A single pass was performed for one third of the plots planned to be harrowed; one third was planned to be treated with two and one third with three passes by harrow. The interval between the passes was one to two weeks depending on the weather conditions.

In 1998, the early harrowing was started on clay (Fig. 3ab) and very fine sand on June 1 and June 9, respectively. The second early pass by harrow was performed on clay (Fig. 3c) and very fine sand on June 9 and June 25, respectively. The late harrowing was started on clay and very fine sand on June 9 and June 25. Because of extraordinary rainy growing season, the second early harrowing was delayed. At the end of June, the camelina had almost reached its maximum height and the flowering started on May 25. Harrowing in this growth stage was considered causing too much damage to the crop and thus, the rest of the planned passes were not performed in 1998.

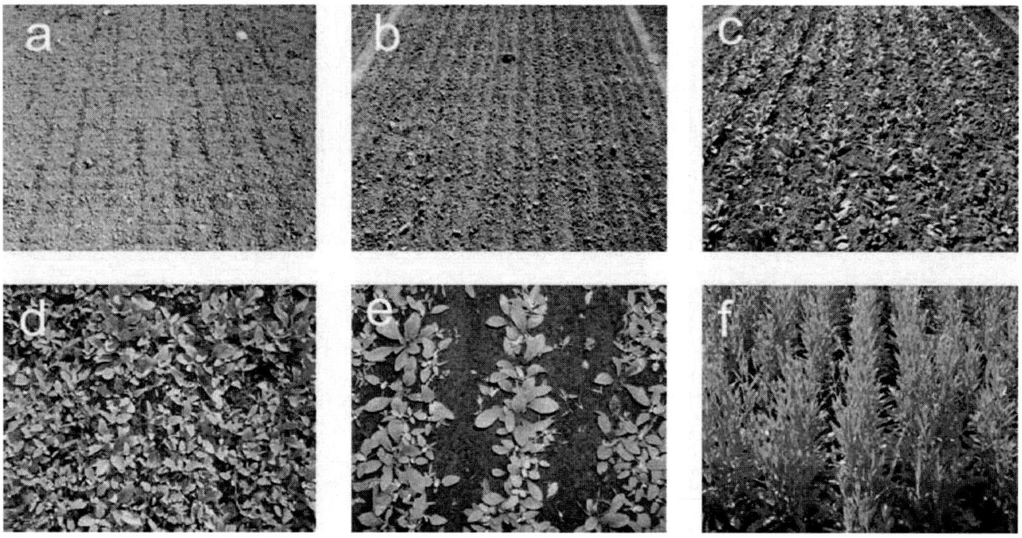

Figure 3. In 1998, the early harrowing was started on June 1, when the camelina growing on clay had one to two leaves. The figures show the situation before (a) and immediately after (b) the first early harrowing treatment. In some of the plots, the harrowing was replicated on June 9, when the camelina had four to six leaves. The figure shows the situation immediately after the second early harrowing (c). In 1999, weed pressure, especially the number of fat hen was high in very fine sand. The effect of the harrowing was, however, relatively good on fat hen in the plots treated with the early harrowing. The figures show an untreated plot (d) and a plot that was harrowed on May 30 and June 7 (e). At the time of the third early harrowing, camelina had almost reached its full height. Figure (f) shows a plot harrowed on May 30, June 7 and June 14.

In 1999, the early harrowing was started on clay and very fine sand on June 3 and May 30, respectively. The second and third early passes by harrow were performed on June 11 and June 17 on clay, respectively, and on June 7 and June 14 on very fine sand, respectively. The late harrowing was started on clay and very fine sand (Fig. 3def) on June 11 and June 7, respectively. The second late pass by harrow was performed on June 17 and June 14 on clay and very fine sand, respectively. The flowering started in June 20 and the third replication of late harrowing was not performed in 1999.

Trifluralin (SuperTreflan) was used for the chemical control. The herbicide was sprayed to the control plots with the amount of 1200 g ha^{-1} prior to the sowing. Immediately after the spray, the plots were harrowed to prevent the degradation of the herbicide. The hand weeded control plots were kept weedless during the growing season. No weed control methods were performed for the unweeded control plots during the experiments.

Observations

In each plot, two sample areas with the size of 100 x 25 cm were systematically selected for the observations. The observations were made during the growing seasons on the weed number and the distribution of weed species. The weeds were counted prior to each harrowing in the control plots and in the plots planned to be harrowed. The number of the weeds was also counted for three to five days after the harrowing. The weed species in the sample areas were also specified at the same time. The number of camelina was counted similar to the weed number, but only in the harrowed plots. At the end of growing seasons, the number of camelina and the weeds within the sample area was counted in all the plots and the weeds were collected for determining the dry weight (105°C, 48 h) of weeds.

In 1998, the plots on clay and very fine sand were harvested on August 26 and September 10, respectively. In 1999, the harvesting was done in clay and very fine sand on August 10 and August 9, respectively. All plants within the plot area were harvested with a small harvester designed for field trials. The harvested seeds were sorted (Kamas Westrup). The moisture of camelina seeds were measured using a grain analysis computer (Dickey John GAC) and the yield was calculated in the corrected water content of 9%.

The Experiment with Winter Annual Cultivar

The winter annual cultivar was sown similar to the annual cultivar on unfertilised very fine sand on September 3, 1998 using the sowing density of 9 kg ha^{-1}. The experimental area was fertilised in the spring of 1999 with 70 kg ha^{-1} N. The plot size was the same as in the experiments with the annual cultivar.

The weed control methods included five different harrowing treatments and the same control treatments that were also performed in the experiments with the annual cultivar. The harrowing treatments were made up of the combinations of the two different timing (early and late) of the harrowing and the number (1x, 2x or 3x) of passes. The early harrowing was started in three fifth of the plots planned to be harrowed on September 25, 1998, when the first leaves of camelina appeared between the cotyledons. The crop over-wintered as a rosette. On May 13, 1999, the second early pass by harrow was performed in two third of the same plots that were also harrowed in the fall 1998. The third early pass was performed for the half of these plots on May 13, 1999.

On May 13, 1999, when the plants had a rosette formed of 6–8 leaves, the first late pass by harrow was performed on two fifth of the plots planned to be harrowed. The second late pass was performed on one half of these plots on May 28, 1999. The flowering of the crop started on June 5, and no harrowing treatments were made after that. The chemical control, hand weeded control and untreated control were performed similar to the control treatments performed in the experiment with annual cultivar. Observations on the number of camelina and weeds, the weed species and the biomass of weeds were also made similar to the experiments with the annual cultivar. The camelina was harvested on August 3, and the yield was measured as in the experiment with the annual cultivar.

Weather Conditions

The weather conditions varied amongst the study years. According to the Finnish Meteorological Institute, the temperature in the summer of 1998, measured from the Helsinki-Vantaa meteorological station (10 km from the experimental area) was near the long term (1960–1990) average, but the summer was rainy. The rainfall exceeded 50 mm, 120 mm and 125 mm in May, June and July, respectively. The rainfall was high, especially in June, when the value was almost three-fold compared with the long term average. The number of totally rainless days in June was only 8. The winter 1998–1999 was clearly warmer and rainier than the long term average. In the summer of 1999, the mean temperature of 19°C for both June and July was high compared with the long term average. The summer was extraordinary rainless; the rainfall exceeded only 11 mm, 25 mm and 14 mm in May, June and July, respectively. The values in May and July were only one third of the long term average and the value in June was approximately one half of the long term average.

Study Design

The study design in the experiments with annual cultivar was a split plot randomised on complete blocks. The main plot factor consisted of the different sowing densities (2 levels) and the sub-plot factor included the different weed control methods (6 or 8 levels in 1998 and 1999, respectively). The whole procedure was replicated in four blocks in two different soil types and in two years. The experiment with winter annual cultivar had a design of randomised complete blocks. The treatments included the different weed control methods (8 levels) replicated in four blocks.

Statistical Tests

A mixed model analysis of variance (ANOVA) with the error structure adjusted for split-plot design was used to test the effects of soil type, year and weed control methods on the yield and the biomass of weeds on annual camelina. Due to the soil type-by-year interaction for the yield, the differences between the different weed control methods were tested within the soil type-by-year using the Bonferroni adjusted pair-wise comparisons of estimated marginal means. Because of the heterogeneity of the variance structure, the data from the number and biomass of weeds on annual camelina was not parametrically tested. A sign test was computed for the differences in weed biomass between the sowing densities. A mixed model ANOVA with the error structure adjusted for a randomised complete block design and Bonferroni adjusted pair-wise comparisons of estimated marginal means were used to test the differences between the means of the yield and number of camelina and the biomass of weeds on winter annual camelina resulted by the different weed control methods. The effects of weed number and biomass on the yield of annual and winter annual camelina were estimated by Pearson correlation. The suitability of the data for the parametrical tests was examined from residual plots. SPSS 12.0 for MS Windows was used for the analysis.

RESULTS

Yield Responses of Camelina to the Different Weed Control Methods

The yield of annual camelina cultivar had an average of 1.6 Mg ha^{-1} over the two years and different soil types (Fig. 4abcd). The yield varied highly over the years and the soil types and the effect of soil type was year-dependent; especially in 1998, the yield in clay was clearly (40%) higher than that in very fine sand (Fig. 4cd; ANOVA, $p<0.01$ for year-by-soil type interaction).

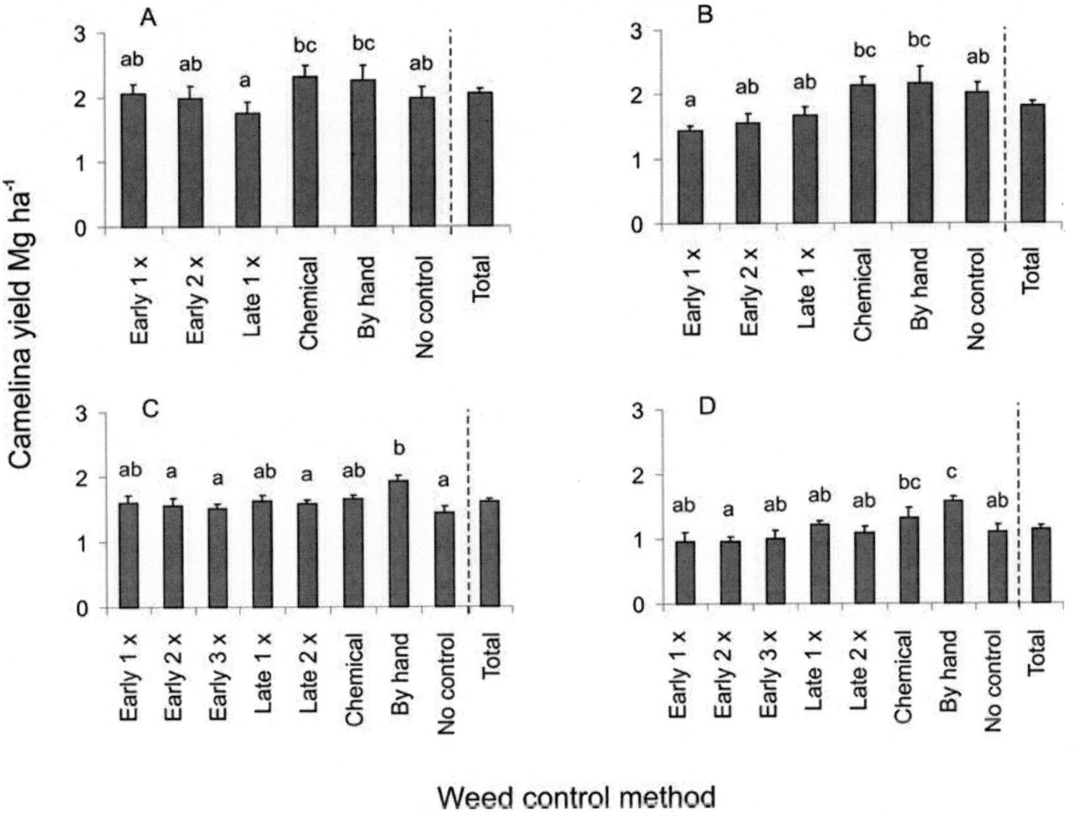

Figure 4.The variation in the yield of annual camelina cultivar affected by the different weed control methods on clay (a) and very fine sand (b) in 1998; and clay (c) and very fine sand (d) in 1999. The bars show the means and error bars indicate standard errors of the means. The difference of the bars marked with the different letters is significant (Bonferroni adjusted multiple pairwise comparisons of estimated marginal means, $p<0.05$). The early harrowing was started at the growth stage of 1–2 leaves of camelina and replicated for once, twice or three times (only in 1999). The late harrowing was started at the growth stage of 4–6 leaves of camelina and replicated for once of twice (only in 1999). Trifluralin (1200 g ha^{-1}) was applied to the chemical control plots. The hand weeded control was kept weedless during the whole duration of the experiment. No weed control was performed for the unweeded control.

Weed control methods clearly affected the yield in 1998 in clay and very fine sand (ANOVA, $p<0.05$ and $p<0.01$, respectively), and in 1999 in clay and very fine sand (ANOVA, $p<0.01$ and $p<0.001$, respectively), but the effects of different harrowing treatments on the yield did not differ from each other (Fig. 4abcd). Most of the harrowing treatments resulted in the yield similar to chemical weed control (Fig. 4abcd). However, hand weeding in very fine sand in 1999 was the only weed control method that managed to increase the yield compared with uncontrolled treatment (Fig. 4d).

Because of the lumpy texture of clayey soil, the emergence of the crop was uneven, especially in the spring of 1999. Thus, the timing of harrowing treatments was not optimal in all parts of the plots. In 1998, the persistence of camelina to the harrowing was very good; the loss of crop plants following the pass by harrow at the 1–2 leaf stage (Fig. 5a, open bars) and 4–6 leaf stage (grey bars) of camelina was only 10–20%. The loss of camelina caused by the harrowing treatments was clearly higher in 1999 compared with that in 1998 (Fig. 5a). This might be a result of the different weather conditions of the growing seasons. In the rainy summer 1998, the recovery of uprooted crop plants was probably good because of the favourable soil water relations.

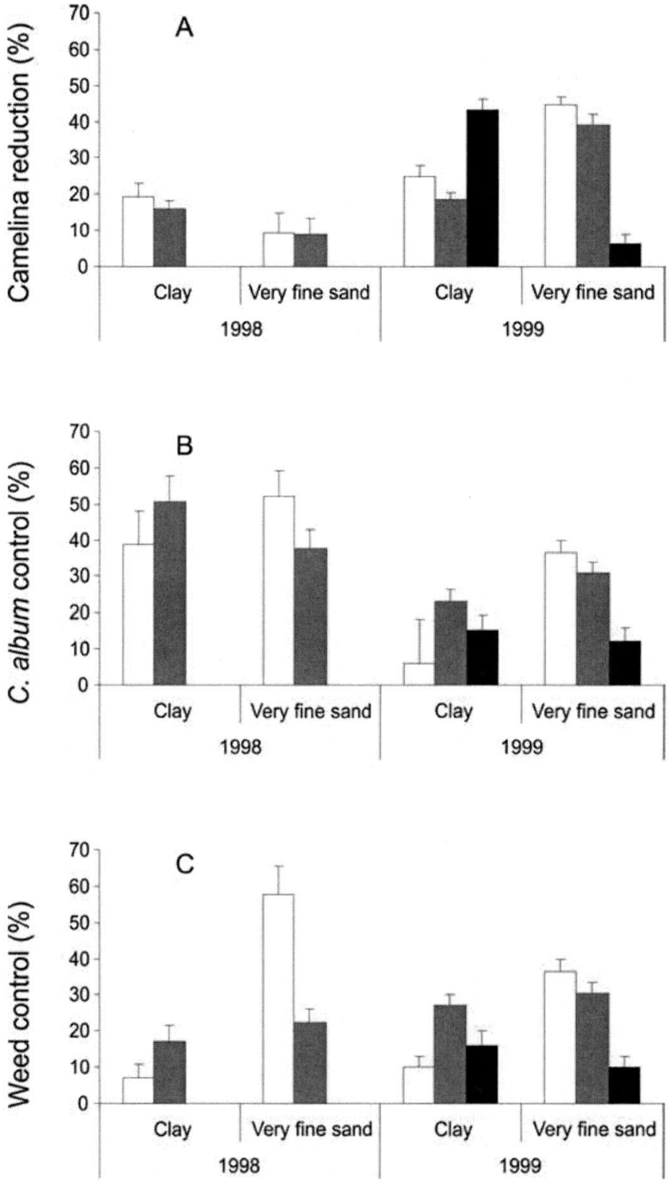

Figure 5. The effect of timing of the harrowing on the reduction (%) in the number of camelina (a), fat hen (*C. album*) (b) and the total of the annual weeds (c) on camelina measured 2–5 days after the pass by harrow and compared with the number of the plants before the pass on clay and very fine sand in 1998 and 1999. White bars represent the means of the first early pass at the growth stage 1–2 leaves of camelina and the cotyledon stage of fat hen. Grey bars represent the means of the second early and the first late pass at the growth stages of 4–6 leaves of camelina and 2–4 leaves of fat hen. Black bars

represent the means of the third early and the second late pass performed when camelina was reaching its full height. Error bars indicate the standard errors of the means.

In 1999, the harrowing treatments made on camelina that had almost reached its full height (Fig. 5a, black bars) reduced the number of crop plants by 50% in clay but only by less than 10% in very fine sand. At the time of harrowing, the uppermost layer of soil was dry and the kinetic energy of the lumpy particles of clay probably damaged camelina more than the loosely structured very fine sand. The recovery of the uprooted crop plants was probably poor in the lumpy structured clay with low content of water available for plants. In all, the harrowing treatments reduced the number of camelina to some extent, but there was no correlation between the yield and number of crop plants (Pearson correlation, $p>0.05$). When growing in low density, camelina is considered being able to compete with other plants by effective branching. In the plots with severe damage caused by harrowing, the loss of crop plants was probably compensated by the branching.

The yield of winter annual cultivar had an average of 1.4 (±0.5, sd) Mg ha^{-1}. Weed control methods had no effect on the yield (ANOVA, $p>0.05$). The persistence of winter annual camelina to the harrowing at the stage of 1–2 leaves (in autumn 1998) was similar to the annual cultivar at the same stage. The number of camelina after the first early pass by harrow was reduced by 20 (±9, sd) % compared with the number before the harrowing. The over wintering of the winter annual cultivar was poor; only 50% of the crop plants managed to survive. Thus, the persistence of crop to the harrowing at the stages of 6–8 leaves and nearly full height of camelina (harrowing treatments performed in spring 1999) was also similar to the annual cultivar harrowed at the rosette stage. The harrowing made at these growth stages reduced the number of crop plants with 11 (±17, sd)% and 15 (±17, sd)%, respectively.

The Biomass and Number of Weeds, Affected by the Control Methods

In 1998, the biomass of weeds on annual camelina cultivar, measured at harvest, had an average of 0.5 and 1.0 Mg ha^{-1} in clay and very fine sand, respectively. In 1999, the corresponding values were 1.5 and 5.0 Mg ha^{-1} in clay and very fine sand, respectively (Fig. 6abcd). The effectiveness of the harrowing treatments was variable; the lowest biomass of weeds was measured at the chemically controlled plots both in clay and very fine sand in 1998 and 1999. There was a weak trend showing that the weed biomass might have been decreased by the increased sowing density. However, the effect was statistically insignificant (Fig. 6abcd).

The most abundant weed species measured on the annual cultivar was fat hen, comprising an average of 53% and 47% of all the weeds in clay and very fine sand in 1998, respectively, and 55% and 99% of the weeds in clay and very fine sand in 1999, respectively. The other weed species abundant on the annual cultivar were chickweed (*Stellaria media*), field pansy (*Viola arvensis*), shepherd's purse (*Capsella bursa-pastoris*), scentless mayweed the brassica (*Brassica spp.*) and knot grasses (*Polygonum spp.*).

During the first early pass by harrow (at the stage of 1-2 leaves of camelina), fat hen was at the cotyledon stage. During the 4-6 leaf stage of camelina, fat hen had 2-4 leaves. The effect of harrowing on the fat hen was good at these early growth stages; the harrowing controlled nearly one half of the fat hen population in 1998 (Fig. 5b). At the growth stage of 2–4 leaves of fat hen the effect of harrowing was poor; harrowing controlled less than 25% of the fat hen population at this stage (Fig. 5b). The effect of harrowing on the annual weeds other than fat hen was poor in clay (Fig. 5c).

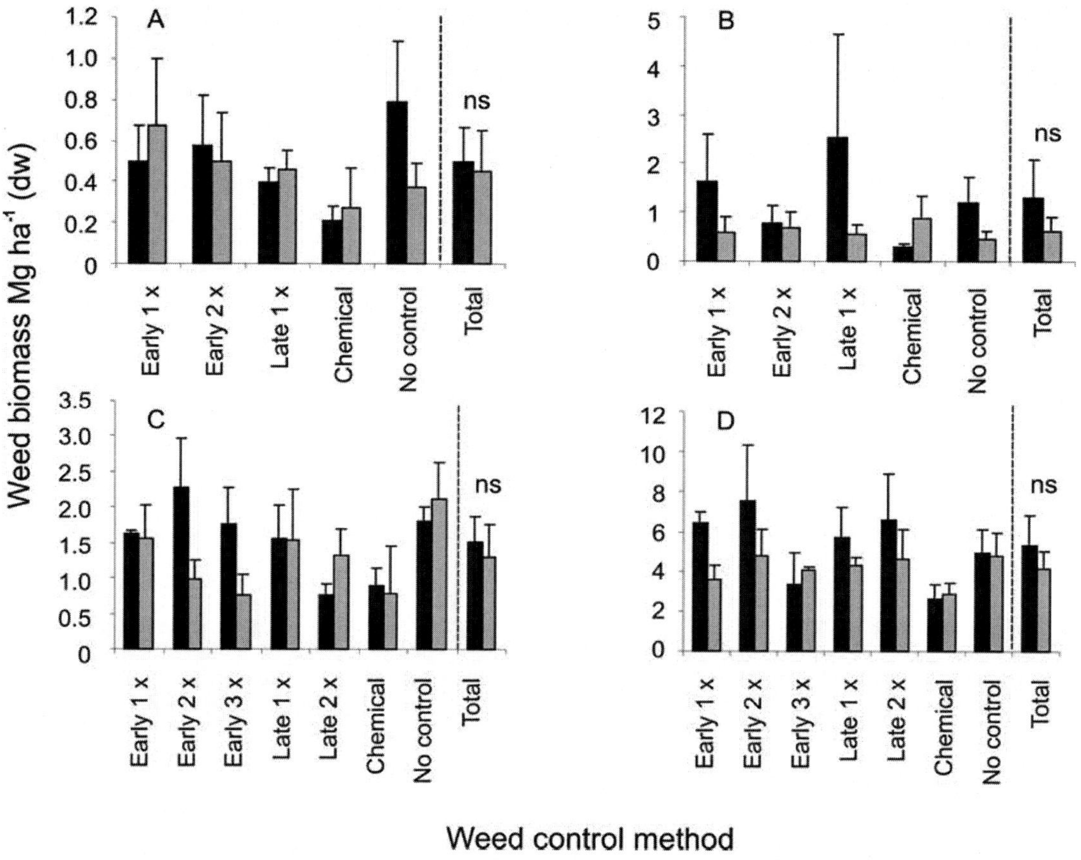

Figure 6.The biomass of weeds on annual camelina cultivar affected by the different weed control methods on clay (a) and very fine sand (b) in 1998; and clay (c) and very fine sand (d) in 1999. The bars indicate the mean biomass of the weeds on camelina with the normal (black bars) and high (grey bars) sowing density. Error bars indicate the standard error of the mean. The letters ns above the bars indicate that the sowing density had no effect on the biomass of weeds (Sign test, $p>0.05$). The early harrowing was started at the growth stage of 1–2 leaves of camelina and replicated for once, twice or three times (only in 1999). The late harrowing was started at the growth stage of 4–6 leaves of camelina and replicated for once of twice (only in 1999). Trifluralin (1200 g ha^{-1}) was applied to the chemical control plots. The hand weeded control was kept weedless during the whole duration of the experiment. No weed control was performed for the unweeded control. Note the different scales in the Y-axis.

The selectivity of harrowing treatments seemed to be good in 1998 (Fig. 5abc); the harrowing treatments controlled effectively the fat hen, with the relatively low losses of camelina. The selectivity was, however, dependent on the year; the harrowing performed in 1998 was more selective than that in 1999 (Fig. 5abc). The selectivity of the harrowing treatments, especially of those made in clay on camelina reaching its full height, was low.

The biomass of weeds on winter annual cultivar, measured at harvest had an average of 3.0 Mg ha^{-1} and was clearly affected by the different weed control methods (Fig. 7). The chemical weed control seemed to result in lower biomass of weeds compared with any of the harrowing treatments. Because of relatively high variation within the different weed control treatments, the multiple pair-wise comparisons did not, however, show any of the control methods being more effective than the others. The most abundant weed on the winter annual cultivar was also the fat hen (64% of all the weeds). Other weed species common on the camelina were field pansy, scentless mayweed, shepherd's purse, scorpion grass (*Myosotis arvensis*), bedstraw (*Galium spp.*) and hempnettle (*Galeopsis bifida*).

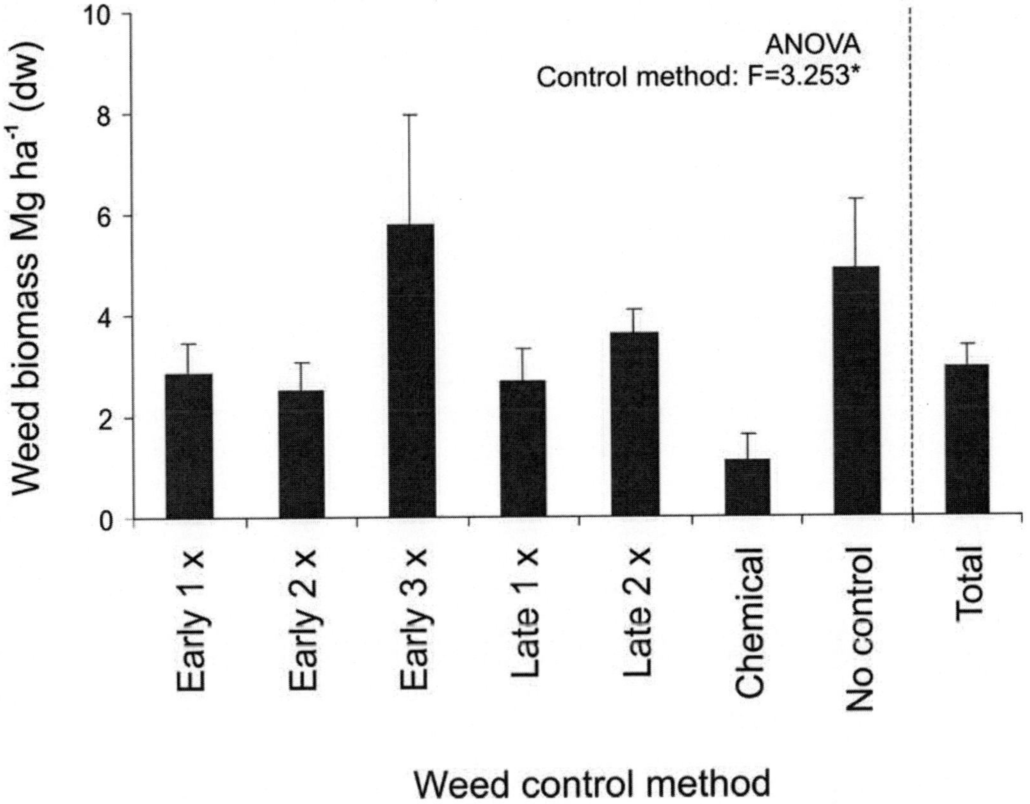

Figure 7. The effect of weed control methods on the biomass of weeds on winter annual camelina. Bars represent the means and the error bars indicate the standard errors of the means. ANOVA results show that the differences between the means was significant (an asterisk, *, means that $p<0.5$).

During the first early pass by harrow (at the stage of 1–2 leaves of camelina, in the autumn 1998), the weeds were at the cotyledon stage. The growth of weeds, especially of scentless mayweed and shepherd's purse, was rapid compared with that of camelina in the spring of 1999. At the time of the first late pass by harrow (at the stage of 6–8 leaves of camelina in spring 1999), the fat hen had 2–8 leaves. Some of the winter annual weeds, such as field pansy, were already flowering at the time of the last harrowing treatment (at the time of camelina reaching its full height).

The first early pass by harrow made at the stage of 1–2 leaves of winter annual camelina controlled 29(±18, sd)% of the weeds measured before and a few days after the harrowing. The effect was good, especially on fat hen. The effect of the harrowing treatments made in spring 1999 was poor; only 11(±12, sd)% and 18(±21, sd)% of the weeds were controlled at the stages of 6–8 leaves and nearly full height of camelina, respectively. The selectivity of the harrowing was also poor in winter annual camelina.

Conclusion

The mechanical control with spring-tine harrow effectively reduced the number of small weeds measured during the growing season. The controlling effect was, however, short-term; new weeds emerged within a few days after a pass with the harrow. The first leaves of camelina form a rosette, which is considered an effective competitive trait suppressing germinating and very small weeds. During the last decade, the farmers' experience has, however, shown that the fast growth during the emergence of camelina is followed by retarded growth at the rosette stage. Controlling the rapidly emerging weeds at this stage is therefore important. The early harrowing replicated two or three times reduced the biomass of weeds on densely sown camelina to the level of the chemical control. The biomass of the weeds, however, had very little or no effect on camelina yield.

Despite the lack of correlation between the yield and biomass of the weeds, weed control is important in view of the yield quality. In addition, uncontrolled weeds reproduce and spread effectively and thus may cause serious problems in the cultivated area. Combining the increased sowing density and early harrowing with several passes by spring-tine harrow could serve as a solution to the weed problems. The two-year study of different weather conditions, however, shows that the success of the harrowing is highly dependent on the year. An intensive harrowing, including perfect timing and several passes, is needed for satisfactory reduction in weed biomass. The selectivity of harrowing may, however, decrease as camelina grows. In this study, the selectivity was poor in clay, especially when performed during the rainless growing season of 1999 on camelina reaching its full height. This indicates that some other options for controlling weeds with camelina should also be tested to achieve a reasonable level of weed control, especially under the conditions that do not favour the harrowing. Currently, cultural methods such as the high sowing density supporting the competitive ability of camelina are highly recommended for the cultivation. The role of crop rotation is also essential for preventing weed problems; weed control should be carefully taken to account in each of the other steps in crop rotation.

Acknowledgements

My sincere thanks are extended to Dr. Riikka Rajalahti and Prof. Eija Pehu for their support and help in the planning and implementation of the study. I also thank Ms. Karita Larkka for the fruitful discussions. Mr. Markku Tykkyläinen and Ms. Lilia Sarelainen are gratefully acknowledged for their skillful help and technical assistance in the field and laboratory.

REFERENCES

Baumann, D. (1992) Mechanical weed control with spring-tine harrows (weed harrows) in row crops. In: *Proceedings IXémeColloque International sur la biologie des Mauvaises Herbes.* Dijon, France; 123-128.

Bernardo, A., Howard-Hildige, R., O'Connell, A., Nichol, R., Ryan, J., Rice, B., Roche, E., & Leahy, J. (2003) Camelina oil as a fuel for diesel transport engines. *Industrial Crops and Products, 17,* 191-197.

Büchsenschütz-Nothdurft, A., Schuster, A. & Friedt, W. (1992) Breeding for modified fatty acid composition via experimental mutagenesis in *Camelina sativa* (L.) Crtz. *Industrial Crops and Products, 7,* 291-295.

Bonjean, A. & Le Geoffic, F. (1999) La Camelinae – *Camelina sativa* (L.) Crantz: Une opportunité pour l'agriculture et l'industrie europ'eennes. *OCL – Oleagineux Corps gras Lipides, 6,* 28-34.

Dönmez, E. & Belli, O. 2007. Urartian plant cultivation at Yoncatepe (Van), eastern Turkey. *Economic Botany, 61,* 290-298.

Duerinckx, K., Mouazen, A., Anthonis, J. & Ramon, H. (2005) Effects of spring-tine settings and operational conditions on the mechanical performance of a weed harrow tine. *Biosystems Engineering, 91,* 21-34.

Fågelfors, H. (1984) Nyttiga ogräs? En kulturhistorisk översikt. Del 5. Lantmannen 6, 28.

Fogelberg, F. & Dock Gustavsson, A-M. (1998) Resistance against uprooting in carrots (*Daucus carota*) and annual weeds: A basis for selective mechanical weed control. *Weed Research, 38,* 183–190

Frame, D., Palmer, M. & Peterson, B. (2007) Use of *Camelina sativa* in the diets of young turkeys. *Journal of Applied Poultry Research, 16,* 381-386.

Francis, C.M. & Campbell, M.C. (2003) *New High quality oil seed crops for temperate and tropical Australia. A report for the Rural Industries Research and Development Corporation.* RIRDC Publication No 03/045; 36.

Fröhlich, A. & Rice, B. (2003) Evaluation of *Camelina sativa* oil as a feedstock for biodiesel production. *Industrial Crops and Products, 21,* 25-31.

Ghersa, C. & Roush, M. (1993) Searching for solutions to weed problems. *BioScience 43,* 104-109.

Gugel, R. & Falk, K. (2006) Agronomic and seed quality evaluation of *Camelina sativa* in western Canada. *Canadian Journal of Plant Science, 86,* 1047-1058.

Hansen, P., Rasmussen, J., Holst, N. & Andreasen, C. (2007) Tolerance of four spring barley (Hordeum vulgare) varieties to weed harrowing. *Weed Research, 47,* 241-251.

Henderson, A., Hallett, R. & Soroka, J. (2004) Prefeeding behaviour of the crucifer flea beetle, *Phyllotreta cruciferae,* on host and nonhost crucifers. *Journal of Insect Behavior, 17,* 17-39.

Karvonen, H., Aro, A., Tapola, N., Salminen, I., Uusitupa, M. & Sarkkinen, E. (2002) Effect of α-linolenic acid-rich *Camelina sativa* oil on serum fatty acid composition and serum lipids in hypercholesterolemic subjects. *Metabolism: Clinical and Experimental, 51,* 1253-1260.

Koch, W. (1964) Unkrautbekämpfung durg eggn, hachen und meisseln im Getreide I Wirkungsweiseund einsatzpunkt van Egge, Hagge und Bodenmeissel. *Zeitschrift für Acker- und Pflanzenbau, 120*, 369-382.

Kouwenhoven, J. (1997) Intra-row mechanical weed control – possibilities and problems. *Soil and Tillage Research, 41*, 87-104.

Kurstjens, D. & Kropff, M. (2001) The impact of uprooting and soil-covering on the effectiveness of weed harrowing. *Weed Research, 41*, 211-228.

Kurstjens, D., Perdok, U. & Goense, D. (1999) Selective uprooting by weed harrowing on sandy soils. *Weed Research, 40*, 431-447.

Lawson, H. & Topham, P. (1985) Competition between annual weeds and vining peas grown at a range of population densities: effects on the weeds. *Weed Research, 25*, 221-229.

Lötjönen, T. 1995. *Kemikaaliton rikkakasvien torjunta riviviljelyssä. Torjuntalaitteisto pienimittakaavaiseen vihannesviljelyyn.* MTT:n Tiedote 17; Agri-Food Research in Finland, Jokioinen, Finland; 38.

Moloney, A., Woods, V. & Crowley, J. (1998) A note on the nutritive value of camelina meal for beef cattle. *Irish Journal of Agricultural and Food Research, 37*, 243-247.

Müller, M., Ordon, F. & Friedt, W. (1999) Perspektiven einer leistungsfahigen getreide- und olsaatenproduktion ackerbaulichen grenzstandorten. *Zeitschrift fur Kulturtechnik und Landentwicklung, 40,* 275-281.

Pachagounder, P., Lamb, R. & Bodnaryk, R. (1998) Resistance to the flea beetle *Phyllotreta cruciferae* (Coleoptera: Chrysomelidae) in false flax, *Camelina sativa* (Brassicaceae). *Canadian Entomologist, 130*, 235-240.

Peiretti, P., Mussa, P., Meineri, G. & Perona, G. (2007) Apparent digestibility of mixed feed with increasing levels of false flax (*Camelina sativa* L.) seeds in rabbit diets. *Journal of Food, Agriculture and Environment, 5,* 85-88.

Plessers, A., McGregor, W., Carson, R & Nakoneshny, W. (1962) Species trials with oilseed pants II Camelina. *Canadian Jornal of Plant Science, 42*, 452-459.

Prankl, H. & Rathbauer, J. (1997). Technical performance of vegetable oil methyl ester with a high iodine number. *The Liquid Biofuels Newsletter, 10*, 3-5.

Pullen, D. & Cowell, P. (1997) An evaluation of the performance of mechanical weeding mechanism for use in high speed inter-row weeding of arable crops. *Journal of Agriculture Engineering and Research, 67,* 27-34.

Putnam, D., Budin, J., Field, L. & Breene, W. (1993) Camelina: a promising low-input oilseed. In: Janick, J. & Simon, J. (Eds.), *New Crops* (pp. 314-322). Wiley, New York.

Rasmussen, J. (1990). Selectivity – an important parameter on establishing the optimum techniquefor weed control in growing cereals. Proceedings 7th European Weed Research Society Symposium, Integrated Weed Management in Cereals, Helsinki, Finland. Pp 197-204.

Rasmussen, J. (1991). A model for prediction of yield response in weed harrowing. *Weed Research, 31,* 401-408.

Rasmussen, J. (1992). Testing harrows for mechanical control of annual weeds in agricultural crops. *Weed Research, 32*, 267–274.

Rasmussen, J. & Svenningsen, T. (1995). Selective weed harrowing in cereals. *Biological Agriculture and Horticulture, 12*, 29-46.

Rasmussen, K & Rasmussen, J. (1999). Barley seed vigour and mechanical weed control. *Weed Research, 40*, 219-230.

Rode, J. (2002). Study of autochtron *Camelina sativa* (L.) Crantz in Slovenia. In: Johnson, C. & Chlodmig, F. (Eds.) *Breeding Research on Aromatic and Medicinal Plants* (pp. 313-404) Haworth Press.

Rokka, T., Alén, K., Valaja, J. & Ryhänen, E.-L. (2002)The effect of a *Camelina sativa* enriched diet on the composition and sensory quality of hen eggs. *Food Research International, 35*, 253-256.

Rüther, H. (1957) Untersuchungen über den Anbauwert von Leindotter als Ölfrucht. Z. Landw.*Versuchs- u. Untersuchungswesen, 3*, 571-581.

Rydberg, T. (1991) *Ogräsharvning. Inledande studie av ogräsharvning I stråsäd. Alternative Agriculture no 8*. Research Committee for Alternative Agriculture. Swedish University of Agricultural Sciences, Uppsala, Sweden; 1-23.

Rydberg, T. (1993) Weed harrowing – Driving speed at different stages of development. Swedish *Journal of Agricultural Research, 34*, 107-113.

Ryhänen, E., Perttilä, S., Tupasela, T, Valaja, J., Eriksson, C., Larkka, K. (2007) Effect of *Camelina sativa* expeller cake on performance and meat quality of broilers. *Journal of the Science of Food and Agriculture, 87,* 1489-1494.

Schuster, A. & Friedt, W. 1995. *Camelina sativa*: Old face – new prospects. *Cruciferae-Newsletter, 17*, 6-7.

Sigareva, M., Earle, E. (1999) Camalexin induction in intertribal somatic hybrids between *Camelina sativa* and rapid-cycling *Brassica oleracea. Theoretical and Applied Genetics, 98*, 164-170.

Vanhala, P. (1992) *Rikkakasvien fysikaalinen ja mekaaninen torjunta kasvukauden aikana. MTT:n Tiedote 7*. MTT Agri-food Research in Finland, Jokioinen, Finland;1-68.

Weissmann, F., Paulsen, H.-M., Fischer, K., Matthäus, B., Bauer, M., Pscheidl, M., Vogt-Kaute, W. (2007) Zum Einfluss der Fütterung auf die Schlachtleistung von Broilern aus Ökologischer Mast Fütterungsversuche mit Leindotterpresskuchen im Hinblick auf die Schlachtköperqualität. *Fleischwirtschaft, 87*, 206-210.

Westman, A., Kresovich, S. & Dickson, M. (1999) Regional variation in *Brassica nigra* and other weedy crucifers for disease reaction to *Alternaria brassicicola* and *Xanthomonas campestris* pv. *campestris. Euphytica, 106*, 253-259.

Wilson, B., Wright, K., Butler, R. (1993) The effect of different frequencies of harrowing in the autumn or spring on winter wheat and control of *Stellaria media* (L.) Vill., *Galium aparine* L. and *Brassica napus* L. *Weed Research, 33*, 501-506.

Vollmann, J., Moritz, T., Kargl, C., Baumgartner, S. & Wagentristl, H. (2007) Agronomic evaluation of camelina genotypes selected for seed quality characteristics. *Industrial Crops and Products, 26*, 270-277.

Vollmann, J., Steinkellner, S. & Glauninger, J. (2001) Variation in resistance of Camelina (*Camelina sativa* [L.] Crtz.) to downy mildew (*Peronospora camelinae* Gaüm.). *Journal of Phytopathology, 149*, 129-133.

Zubr, J. (1992) New vegetable oil for food application. *Agro Food Industry Hi-Tech, 4*, 24-26.

Zubr, J. (2003) Qualitative variation of *Camelina sativa* seed from different locations. *Industrial Crops and Products, 17*, 161-169.

In: Weeds: Management, Economic Impacts and Biology
Editor: Rudolph V. Kingely
ISBN: 978-1-60741-010-2

Chapter 4

The Contribution of Arable Weeds to Biodiversity

***D.M.B Parish*[a], *C. Hawes*[b], *S.P. Hoad*[c], *P.P.M. Iannetta*[b] and *G.R. Squire*[b]**

a The Game and Wildlife Conservation Trust, 2 East Adamston Farm Cottages, Muirhead, Dundee, DD2 5QX, UK
b Scottish Crop Research Institute, Invergowrie, Dundee DD2 5DA, UK
c SAC, West Mains Road, Edinburgh EH9 3JG, UK

Abstract

There is increasing recognition in Europe that arable plants have functions beyond simply suppressing the yield of crops. The function best quantified to date is the support they provide to higher trophic groups and to the whole food web of farmland. This food web mediates essential ecological processes such as the element cycling and pollination, and contributes aesthetic qualities through butterflies, birds, rare plants and landscape. The food web has been seriously diminished in the last half-century to the extent that it is now much less than 1% of the total summer biomass in intense arable systems. This paper summarises the evidence for the importance of arable plants to higher trophic groups and discusses the means to regenerate them with least detriment to crop yield. It shows that the field margins and the cultivated area sown with crops have their distinct contributions and should be managed integratively.

Introduction

Weed may be defined as "*a plant whose populations occur naturally and most frequently in environments that are frequently disturbed (by man)*' (Baker 1965, 1974). In this article, we identify climate change and the development of cropping practices by man as key features in the history and evolution of weed species, and define the role and importance of weeds within and around arable land, with reference to Europe and to Britain in particular. The effects of cropping activities on the reductions and other alterations in the diversity of within-field weed communities are summarised. The consequences of these changes in the weed flora for the insect and bird fauna of arable land are also discussed in the context that farmland biodiversity is essential, not incidental, for continued crop production. New developments in arable management, mainly in Britain, that aim to redress the balance between production and biodiversity are described with emphasis on future directions. A greater understanding of weed species will be required if management of the farmed ecosystem is to be sustained. Informed decision-making must balance the long-term economic benefits of greater weed-crop co-existence against short-term costs such as slightly decreased yield (Costanza et al. 1997, Mader et al. 2002).

1. Evolution Of Weeds in Land Under Arable Crop Production

In evolutionary terms, a 'successful weed species' is a plant that produces offspring which are reproductively successful in environments that place plants under a high degree of stress caused by frequent disturbance. Compared to many other plants, weeds grow rapidly through their vegetative phase to flower and complete their life-cycle in a short time. Many of them are self-fertile and employ wind or generalist insects for pollination. They deposit many small seeds that are returned to the soil as a weed reservoir (the '*seedbank*'), where they may become dormant, retaining their viability for many years. This range of adaptations has evolved in response to regular disturbance, since many weed species have ancestors with late time to flowering and out-crossing that relies upon specialist pollinators (Yan et al. 2003, Komeda 2004, Roux et al. 2006). Such archetypal species originate from regions of the world that were arctic *refugia*.

The weed biota of central and northern Europe (including Britain), was established by the colonisation of land exposed in the wake of a retreating ice-cap after the most recent ice-age (Hewitt 1999). This rapid re-colonisation was assisted by human migrations across Europe in the bronze and iron ages and subsequent farming activity (Gray 1897) that encouraged weeds to develop traits facilitating their co-existence with crops. For example, the selection for rapid cycling (that is, early time-to-flowering) in response to crop production and weed control in recent times has been noted for wild oats (*Avena* spp; Imam & Allard 1965) and thalecress (*Arabidopsis thaliana*; Jones 1971), while arable populations of the common weed, *Capsella,* mostly have a short time to flowering and low leaf biomass (Iannetta et al. 2007). The characteristics of in-field weeds in Europe must therefore be considered to have evolved in response to a warming climate and human activity.

Most importantly, weeds are still evolving, adapting to intense selection pressures. The most dramatic example of this capability is the increasing prevalence of weeds resistant to commonly used herbicides, such as glyphosate (Holt & LeBaron 1990, Powles 2008). Selection for such an obvious adaptation is no more ecologically significant than the enrichment of weed populations with other less obvious traits (e.g. time to flowering and fecundity) that may result from non-chemical interventions such as the use of specific crop-types or tillage practices (Hawes et al. 2005, Owen 2008).

Weeds have coexisted with crops for several millennia, therefore, and the last half-century of intense, chemically-aided cropping has not succeeded in obliterating them. Indeed, highly aggressive species are still present, causing major economic problems, while the main casualties have been the neutral or beneficial species that constitute much of the botanical diversity of farmland. Managing biodiversity must therefore be integral to a cropping system if the selection pressures that drive weed adaptation are to be reduced and sustainable weed management achieved (Kogan 1998).

2. The Effect of Lost in-Field Weed Diversity and the Need to Manage the Crop-Weed Balance

The weeds of cropped land are mostly wild species that respond to opportunities in the form of germination sites and available resource for plant growth provided by cultivation of soil. The weeds bring problems for agriculture: they compete with the crop, introduce impurities, some poisonous, into the harvest and compromise the efficiency of harvesting methods and machinery. The main aim of weed-management in agriculture has been to reduce weed population density and mass of weeds, so as to lessen or even remove such adverse effects. In Europe, the transition from mainly spring-sown to autumn-sown (winter) cereals and the increased use of chemical herbicides (both since the 1960s) have brought about large and widespread reductions in the weed flora and associated biodiversity (Chamberlain et al. 2000, Robinson & Sutherland 2002, Marshall et al. 2003). Despite appearances, the habitat can still support a wide range of wildlife. Over 130 plants and 550 arthropods were recorded during one long-term study in Sussex, UK (Potts 1991).

How much do weeds currently contribute to primary production - the uptake of carbon and its conversion to biomass? The annual primary production of crops in north temperate lands is 15-30 t ha^{-1} (Monteith 1977), while the corresponding production by in-field weeds, measured in the largest survey of such in the UK, is around 0.03 t ha^{-1} or 1% of the in-field primary production by break crops such as oilseed rape, and typically ten times less than this in intensely managed cereals (Heard et al. 2003, Bohan et al. 2005, Hawes et al. 2009). Such a small weed mass has little effect on the yield of crops. Indeed, large relative changes (e.g. two-fold) in these low weed populations have a negligible influence on yield in cereal rotations (Young et al. 2001). In general, intense cereal management has achieved the aim of reducing weeds.

The reduction of weeds in European cropping has not been without other effect. There is increasing evidence, and appreciation, that the weed flora has a range of qualities that are beneficial to the farmed ecosystem and to human needs. The weeds have among their number many rare plants that are considered in some countries as part of the national, historical flora that is in rapid decline (Sotherton & Self 2000, Preston et al. 2002). Also, among the arable weeds are plants that exhibit medicinal and pharmaceutical properties that have not been fully investigated. They are themselves part of the biodiversity, but they also support a level of animal and microbial biodiversity far greater than is supported by crops alone (Sotherton & Self 2000, Hawes et al. 2005). They also, especially if allowed to grow in the margins of the cropped field, restrict rainwater-runoff and take up nutrients that would otherwise enter water courses and contribute to eutrophication and loss of non-renewable resources. In total, arable crop lands if managed sympathetically could make a major contribution to the conservation of biodiversity, food webs and rare species in Europe.

Arable systems therefore need to be managed around an optimum state in which several outcomes are satisfied, not necessarily always in the same field, but over time and within the landscape. A particular balance needs to be achieved between the undoubted negative effects of weeds on yield and the range of positive effects on the ecosystem. The means to reach this balance are not straightforward, since it needs an understanding and manipulation of complex mixed plant populations and even more complex food webs.

3. The Nature and Status of Weeds Within and Around Cropped Fields

The vegetation associated with arable farming comprises the in-field weeds of the cropped area, sown or managed margins or strips of usually perennial species, the boundary vegetation of perennials and hedges, assorted woodland and parcels of unused land. The balance between these types differs greatly across Europe, but generally the in-field weed vegetation occupies the greatest area, and in some regions is almost the sole vegetation apart from the crops where fields abut against each other without boundaries. The in-field vegetation also usually contains the most plant species in the arable landscape. In the largest UK-wide survey in recent times, around 260 species were recorded during sampling (which was not exhaustive) in 65 fields of winter oilseed rape (Hawes et al. 2009, Squire et al. 2009) and probably the total in-field flora in the UK is well above 300 species, consisting of annuals, biennials, and perennial grasses and broadleaves, together with shrub and tree seedlings that habitually ingress from the boundaries. The most abundant species are annuals and biennials, except where grass leys occur in the cropping sequence for two or more years when perennials may become prevalent.

The undesirable weeds are mainly of two types: those that are poisonous or anti-nutritional and those that reduce yield by competing with the crop. The poisonous and anti-nutritional have been largely reduced or removed in intense cereal rotations, though persist in less developed agriculture in parts of Europe. Species such as henbane (*Hyoscyamus niger* L.) and corncockle (*Agrostemma githago* L.) are now very rare in fields managed intensely for food and fodder, while others such as ragwort (*Senecio jacobeae* L.) remain in poorly managed pasture and along waysides. The main weed burden is represented, typically, by 5 to 10 species at any time. In the early part of the 20th century, the main species included the thistles (species of *Cirsium*) and docks (species of *Rumex*), which were subject in the UK to enforcement orders if not controlled by the land owner. Today, these are mainly found in, again, poorly managed pasture and in waste ground and waysides. The economically damaging weeds today are those that compete with the main cereals, and include grasses such as wild oat (species of *Avena*), brome (species of *Bromus*), couch (*Elytrigia repens* (L.) Desv. ex Nevski) and blackgrass (*Alopecurus myosuroides* Huds.) and certain smothering broadleaf species such as cleavers (*Galium aparine* L.) and chickweed (*Stellaria media* (L) Vill.). The biennials and perennials tend not to last beyond a year in cultivated land, but most of the remaining 200 or more annuals and winter annuals, the latter germinating in autumn and flowering the following spring or summer, tend to be reduced by the broad spectrum herbicides that are increasingly used in cereal rotations (Marshall et al. 2003).

Decline and Replenishment

Declines in the population size of seedbank species occur because emerged plants are killed or suffer reduced seeding, while the seedbank is depleted continuously through seeds dying, being attacked by pathogens or being eaten. If an arable seedbank is not replenished, the abundance of all its species combined decreases typically by about 50% a year, but around that figure, individual species vary greatly (Brenchley & Warington 1933, Roberts 1958, 1962, Hawes et al. 2005, Squire et al. 2005). However, the declines in the seedbank have not occurred universally, even in regions where crop management is generally intense, such that a defining feature of seedbank surveys in the UK since around 1910 (Brenchley 1920) has been a very wide range in the seedbank abundance among a sample of fields, usually from 100-1000 m^{-2} (seeds) up to 10,000-100,000 m^{-2}. Low seedbanks were found in the early 20th century as a result of intense, mechanical weed control and are not a feature only of the chemical control of the past quarter century. By the end of the 20th century, the geometric mean of seedbanks in the largest ever survey in the UK (Champion et al. 2003) were 2250 m^{-2} (seeds to a depth of 0.15 m) compared to the geometric mean of 6680 m^{-2} estimated from all previous surveys between 1910 and 2000 (Squire et al. 2003). Moreover, the declines can to a degree be reversed over around 5 years by varying the intensity of weed control within a field (Squire et al. 2000), though whether a species complement can be reinstated is uncertain.

Nevertheless, the rate of replenishment has been less than the rate of decline over a wide enough area to cause a reduction of weed species in the past 50 years, as shown by national surveys on a 10 km square recording grid (Preston et al. 2002). Most species that are declining rapidly are either neutral, have conservation value as rare species, or have beneficial properties that support the food web. There are remarkably few consistent or systematic estimates of weed mass, population density or number of species per unit area (Marshall et al. 2003, Hawes et al. 2005), but there is no question that many rare species are under threat and some are being reduced to extinction, such as the corncockle and thorow-wax (*Bupleurum rotundifolium* L.).

Units of Biodiversity

Biodiversity of the weed flora is typically accounted by species, a currency that provides information on relative rarity, and which may be useful for assigning conservation status of plants in this habitat, but is of little use in demonstrating the ecological effects of the arable flora. There has been little direct measurement of actual traits, such as architecture and nutritional quality, of weeds in the field. To date, approaches to defining the functional nature of weeds have inferred traits from the general characteristics of the species (Storkey 2006, Hawes et al. 2009). For example, the knotweed (*Polygonum aviculare* L.) can be classed as an annual, spring-germinating, indeterminate, spreading, within-canopy plant type, and shepherd's purse (*Capsella bursa-pastoris* (L.) Medik.) as an annual, spring- and autumn germinating, determinate, compact, below-canopy type. Using such classifications, augmented by knowledge of the dietary intake by invertebrates and birds, would make it possible to define the potential of a seedbank to provide both a weed burden and a source of food for these groups at various times of the year. However, little is known about the specific dietary requirements of weed-eating farmland invertebrates in the UK.

Moreover, classifications based on inferred function may be misleading. The functional traits are not adequately known for many weed species, which as a whole are poorly researched (Hawes et al. 2005). If the variation of ecologically functional traits within a species is very wide, as we consider it is, then a species may contain more than one functional type. Common and widespread species such as *Capsella bursa-pastoris* appear to contain, in traits such as time to flowering, as great a range of variation as occurs in most of the annual weed flora (Iannetta et al. 2007). Nevertheless, as is described in the next section, functional (as distinct from purely species-based) approaches are essential for quantifying food webs based on the seedbank and marginal flora.

At the scale of the community, arable seedbanks possess characteristics typical of much other vegetation in having regular species accumulation and rank abundance properties (Magurran 1988) which condense the abundances and distributions of the species in a field into a descriptor (curve) with a few parameters that allow comparison between fields and agronomic treatments. Species accumulation and rank abundance are community-scale descriptors that may be taken to indicate the impact of management on the whole seedbank (Squire et al. 2000). Rank abundance curves can be constructed by using existing data in, for instance, Brenchley & Warington (1933) and Milton (1943). It should be possible therefore to compare the effects of changed management on the seedbank through such curves independently of the complement of species, though to our knowledge such a comparison has not yet been done.

4. In Field Weeds are Essential for Robust, Sustainable Invertebrate Food Webs

Arable food webs are based on primary production by the crop and the weed flora. These plant communities provide a range of different resources for primary consumers (herbivores, pollinators and detritivores), which in turn support an array of natural enemies from host-specific parasitoids to wide-ranging generalist predators. Weeds, and dicotyledon weeds in particular, represent a highly valuable resource to primary consumers within arable fields, supporting up to ten times the biomass of herbivores per unit plant mass compared to crop plants (Hawes, unpublished data). Diversity in this weed layer results in increased diversity of herbivore consumers (Marshall et al. 2003), and some studies have shown that this effect is propagated through the food web to the third trophic level (e.g. Hawes et al. 2003, Taylor et al. 2006).

Farm management practices that reduce the abundance and diversity of weeds, and the strength of their interactions, are therefore likely to have knock-on effects to higher trophic levels, resulting in an overall reduction in arable biodiversity. The extent to which these changes occur depends on the strength of the interactions between resource and consumer groups and the degree of competition for resource. These processes are thought to determine the level of functional diversity which in turn influences the stability or resilience of the system (Root 1973, Kinzig et al. 2002). Declining diversity is generally considered detrimental to ecosystem function and sustainability (Tilman et al. 1997, Loreau 1998, Hector 2001, Scherer-Lorenzen et al. 2003). However, much experimental work relies heavily on taxonomic assessments of diversity (e.g. species number) which fails to account for the differences within and between species that relate directly to the functioning of the ecosystem (Diaz & Cabido 2001, Diaz et al. 2003, Naeem & Wright 2003). By adopting a functional approach to arable diversity, where species are classified according to functional traits, rather than relying on traditional taxonomic definitions, a better assessment of system responses to change might be made (Hawes et al. 2005, Storkey 2006). A functional approach not only allows direct evaluation of the effects of change on the functioning of the system, but also avoids the assumptions that all taxonomic units are functionally equivalent and that there is no redundancy in the system. Because these assumptions are unlikely to hold true in many systems, and vary in degree between systems, no universal relations between taxonomic diversity and system function have been established. A more general theory might therefore result from analysis based on functional groupings of organisms.

Functional groups may be defined according to a set of eco-physiological traits that relate to the process of energy or nutrient cycling and are correlated with productivity, though other criteria could be applied, depending on which system process is of interest (Lavorel 2001). Many of the species inhabiting arable land share a range of functional life history or behavioural characteristics that allow persistence in these regularly disturbed systems. Classification into functional groups therefore requires a more refined assessment of the differences between them. This can be done on the basis of physiological or morphological characteristics that determine resource capture and allocation (an organism's ability to acquire resource, grow and reproduce) and resource transfer (cycling of energy, carbon and nitrogen through the food web). These functional traits relate to the system-level processes of nutrient and energy utilisation and flow, and can therefore be used to assess the response of the system to changes in management.

Surveys of arable weeds and insects across the UK (Firbank et al. 2003) have shown that the strongest interactions within the arable food web tend to centre on the omnivores and form two main interacting groups: one between detritivores and generalist predators, and one between leaf chewers, sap feeders and parasitoids, the omnivores acting as both predator and prey in each case (Hawes et al. 2009). Some herbivore functional groups respond positively to plant biomass, but others are more related to the abundance of their predators than by the amount of available plant resource. Both top-down and bottom-up regulation of populations have been described in other terrestrial systems, e.g. grassland habitats (Siemann et al. 1998; Knops et al. 1999). The effect of this top-down control by natural enemies, which may not themselves be directly influenced by the plant layer, will tend to buffer the herbivore community from small changes in the availability of their resource.

The relative abundance of different functional groups inhabiting the cropped area of arable fields throughout the UK has been shown to be determined primarily by the season of sowing (autumn or spring), the type of crop that is sown (species, variety) and, to a lesser extent, the weed management regime imposed on that crop (Bohan et al. 2005, Hawes et al. 2003, 2009). This significant influence of sowing season has major implications for the functional composition of plant and invertebrate assemblages in arable systems given the trend from spring cropping to a greater proportion of autumn-sown crops in the UK since the 1970s (Chancellor 1985). Relative to spring-sown break crops, autumn-sown crops shift the composition of arable plants towards a monocot dominated community with a greater proportion of the longer-lived, perennial functional types (Hawes et al. 2009). Invertebrate community composition is likely to be similarly affected, since winter crops tend to favour autumn-breeding species and may also provide suitable conditions for species that overwinter as adults or larvae rather than as eggs.

It is argued here that a balance of different functional types of organisms in the arable system is necessary to maintain the long-term functioning and sustainability of that system. Evidence of this is now strong for the organisms that mediate processes such as comminution of detritus, predation and pollination. The balance of functional types is dependent on the presence of an adequate (and probably small compared to the crop) cover of annual weeds within the cropped area, as well as sufficient marginal vegetation to support viable populations of invertebrates and other organisms during periods of within-field disturbance and low weed cover. Tolerance of some weed cover within fields is therefore essential for the maintenance of the within-field functions, including decomposition, soil nutrient retention and control of pest and disease populations. Although the within-field arable food web is central to the provision of these ecological functions, and is therefore important in its own right, it also provides the necessary resources to support consumer populations at greater spatial and temporal scales. The challenge is to define the mass and composition of the weed flora that achieves these ecological functions with minimum agronomic impacts.

5. Farmland Birds as Positive and Negative Indicators of Sustainable Farming Practice

The most studied group inhabiting farmland (at least in the UK) is the birds. Farmland birds have declined in range and number recently across Western Europe (Tucker & Heath 1994, Pain & Pienkowski 1997). In the UK, farmland birds have declined more than species associated with other habitats (Fuller et al. 1995) and farmland specialists have declined more than generalists (Gregory et al. 2004). The loss of species has been greatest in western, grassland dominated, regions of the UK (Chamberlain & Fuller 2000) such that the most recent atlas of UK breeding birds revealed that farmland bird abundance was greatest in arable and mixed farming areas of eastern Britain (Gibbons et al. 1993), highlighting the importance of the arable environment for these species.

The ultimate causes of decline are known to a greater or lesser degree for most farmland birds in the UK (Newton 2004) and for the majority of cases, a reduction in the availability of food at some point during the annual cycle has been implicated, either impacting on its own (e.g. yellowhammer *Emberiza citrinella*) or in concert with other factors (e.g. loss of habitat, lapwing *Vanellus vanellus*), most ultimately linked to changing agricultural practices (Moreby & Southway 1999, Chamberlain et al. 2000, Newton 2004). The main vehicles capable of delivering changes to farmland biodiversity in Europe are agri-environment schemes. These provide grant-aid to farmers in return for wildlife-friendly management, much of which is targeted at farmland birds. There is some evidence suggesting that these measures may have benefitted some species of birds (Aebischer et al. 2000, Swash et al. 2000, Kleijn & Sutherland 2003, Bradbury et al. 2004), but many such as linnet (*Carduelis cannabina*) and sky lark (*Alauda arvensis*) are still in decline nationally (Baillie et al. 2007), suggesting that either the scale of habitat manipulation or the nature of the specific prescriptions are falling short of providing all that is required to halt the declines.

Within these schemes currently, there are few prescriptions applicable to the whole arable crop: most apply to boundary and non-crop habitats. The margins are often well used by farmland birds (e.g. yellowhammers foraging in grass margins, Perkins et al. 2002) and can have positive impacts on the target species. For example, conservation headlands, whereby the use of pesticides in cereal margins is restricted thereby increasing weed and invertebrate populations, can increase grey partridge (*Perdix perdix*) breeding success and reverse population declines (Rands 1985, Sotherton 1991, Aebischer et al. 2000). However, the potential is considerable for in-field manipulations to be accessible to many more birds, not only because of the larger area of manipulated habitat, but also because they would directly benefit species preferring open habitats, like the sky lark.

Given that the loss of weeds as an energy source (food), and also as a crucial link in the food chain, has been implicated in the decline of many species of farmland birds, one in-field manipulation worthy of investigation is the tolerance or encouragement of a limited community of arable weeds within the crop itself (Marshall 2004). These would provide a rich source of food directly (Holland et al. 2006), increase the value to birds of any stubble or fallow retained after the crop is harvested (e.g. Moorcroft et al. 2002) and - as argued earlier in this paper - provide a haven for invertebrates that themselves are an important source of food for many birds (Holland et al. 2006).

This prompts the question is there evidence that birds would use fields, as opposed to field edges, to exploit the additional resources? Many species of birds use arable fields in winter, with seed-eating species mostly associated with seed-rich habitats like stubbles (Wilson et al. 1996, Robinson & Sutherland 1999) and invertebrate-eating species particularly attracted to permanent pasture (Tucker 1992, Wilson et al. 1996). It is not clear (i.e. the evidence is not conclusive) for most species whether they would use resources away from the field edge. Sky larks probably would as they often avoid the field edge during the winter (Robinson & Sutherland 1999). In contrast, wintering yellowhammers show the reverse trend, avoiding field centres, although this may reflect food availability as seed density has been shown to decline with distance from the field edge (Robinson & Sutherland 1999) and yellowhammers will move further away from the field boundary as seeds become depleted (Robinson & Sutherland 1997). However, many farmland passerines seem to remain close to cover at the field edge as a means of minimizing predation risk (e.g. Whittingham & Evans 2004) and there is likely to be a complex interaction between food availability and predation risk that ultimately determines the distance to the field edge selected by foraging birds. However, it does seem likely that increasing the diversity and availability of seeds in any retained stubble fields would almost certainly benefit seed-eating birds.

It is also not clear if increased weed abundance within the crop would ultimately increase the use of fields in winter by invertebrate-feeding bird species. Their documented preference for pasture has been explained by the abundance of invertebrates in such habitats (Tucker 1992). Many lapwings, for example, are found on arable fields in the winter, but it is not clear if this reflects preference or simply the number of birds wintering in eastern Britain (Sheldon et al. 2004). Lapwings are known to avoid the field edge (Thompson & Barnard 1983, Barnard & Thompson 1985) and so would probably be able to exploit any additional resources provided. The general avoidance of the field centre by passerines presumably holds for invertebrate feeders on grassland, but to our knowledge this has not been explored.

Many farmland birds forage within arable crops during the breeding season (e.g. grey partridge, Green 1984; yellowhammer, Stoate et al. 1998; corn bunting (*Emberiza calandra*), Brickle et al. 2000), although the precise selection of foraging locations within fields in relation to the edge during the breeding season is less well known. It seems likely, however, that birds would be more willing to forage further from the field edge at this time of year as cover is provided by crops.

Birds are often regarded as indicators of the state of farmland and their recent population declines have been mirrored in other taxa, as mentioned above. The fact that weeds are of importance to most farmland birds, either as a direct food source or as a medium supporting other food items (invertebrates – see section 4 above), any management prescription that increases weed biomass (especially dicotyledonous weeds) in a way that is accessible to birds has the potential to benefit the entire farmland bird community, as well as the weed community and various food webs that depend on it (sections 3 & 4). The challenge for researchers is to find a means of increasing in-field weed communities in a way that maximizes exploitable resource provision and minimizes agronomic impacts.

6. Conclusions, Recommendations and Research Targets

- The arable farmed habitat is increasingly recognised as supporting a range of organisms that are variously integral to ecological functioning and aesthetically valuable to people. The aesthetic appreciation of farmland habitats and organisms is particularly important in those regions or countries where the arable landscape constitutes a large part of the total landscape.
- Research over the previous two decades has made substantial gains in knowledge of this hitherto under-researched arable habitat, but uncertainties remain in the links between weed functional types and the invertebrates and birds that live in the habitat. There is very little knowledge of the links between weeds and microbial functioning in the soil.
- Agri-environment schemes have tended to concentrate on arable field margins and strips, ignoring the in-field vegetation and the ecological processes that are mediated by the weeds as well as the crops. Yet the in-field, cultivated land occupies the greatest area and relies on a food web *in situ* to mediate functions such as decomposition and pollination.
- There is still considerable scope to design prescriptions based on improving the habitat as a whole: these need not mean diverting large amounts of primary production away from the crop, but should encourage and sustain the arable food web generally and not only species with aesthetic value.
- Any proposals should be founded on a better understanding of the cycling of energy and matter in arable systems, both over time through the rotation and over space in the landscape. We consider the following specific questions to be worthy of urgent investigation: 1) Which species of weeds, or functional types, will maximise biodiversity benefits while minimizing agronomic impacts on? 2) What is the appropriate level of weed abundance or cover that will, again, maximize benefits while minimizing impacts? 3) What is the best way to encourage weeds within the

crop (e.g. direct sowing or changes to crop management) without detriment to farm profit?

ACKNOWLEDGEMENTS

Dave Parish is supported by the Game and Wildlife Conservation Trust. Cathy Hawes, Steve Hoad, Pietro Iannetta and Geoff Squire are supported through the *Sustainable Agriculture - Plants* research programme funded by the Scottish Government. Many thanks to Nick Sotherton, John Holland and Barbara Smith for helpful comments.

REFERENCES

Aebischer, N.J., Green, R.E. & Evans, A.D. (2000). From science to recovery: four case studies of how research has been translated into conservation action in the UK. In: Aebischer, N.J., Evans, A.D., Grice, P.V. & Vickery, J.A. (eds). *Ecology and conservation of lowland farmland birds*. Tring, British Ornithologists Union. Pp. 43-54.

Baker, H.G. (1965). Characteristics and modes of origins of weeds. In, *The Genetics of Colonising Species*. Baker, H.G., Stebbins, G.L. (eds). New York, Academic Press. Pp. 147-72.

Baker, H.G. (1974). The Evolution of Weeds. *Annual Review of Ecology and Systematics* **5**, 1-24.

Baillie, S.R., Marchant, J.H., Crick, H.Q.P., Noble, D.G., Balmer, D.E., Barimore, C., Downie, I.S., Freeman, S.N., Joys, A.C., Leech, D.I., Raven, M.J., Robinson, R.A. & Thewlis, R.M. (2007). *Breeding Birds in the Wider Countryside: their conservation status 2007*. BTO Research Report No. 487. Thetford. BTO. (http://www.bto.org/birdtrends)

Barnard, C.J. & Thompson, D.B.A. (1985). *Gulls and Plovers: the ecology and behaviour of mixed species feeding groups*. London. Croom Helm.

Bohan, D.A., Boffey, C.W.H., Brooks, D.R., Clark, S.J., Dewar, A.M., Firbank, L.G., Haughton, A.J., Hawes, C., Heard, M.S., May, M.J., Osborne, J.L., Perry, J.N., Rothery, P., Roy, D.B., Scott, R.J., Squire, G.R., Woiwod, I.P. & Champion, G.T. (2005). Effects on weed and invertebrate abundance and diversity of herbicide management in genetically modified herbicide-tolerant winter-sown oilseed rape. *Proceedings of the Royal Society B-Biological Sciences* 272, 463-474.

Bradbury, R.B., Browne, S.J., Stevens, D.K. & Aebischer, N.J. (2004). Five-year evaluation of the impact of the Arable Stewardship Pilot Scheme on birds. *Ibis* 146 (suppl. 2), 171-180.

Brenchley W.E. (1920). *Weeds of farmland*. Longmans, Green & Co., London.

Brenchley W.A., Warington K. (1933). The weed seed population of arable soil. II. Influence of crop, soil and methods of cultivation upon the relative abundance of viable seeds. *Journal of Ecology* 21, 103-127.

Chamberlain, D.E. & Fuller, R.J. (2000). Local extinctions and changes in species richness of lowland farmland birds in England and Wales in relation to recent changes in agricultural land-use. *Agriculture Ecosystems and Environment* 78, 1-17.

Chamberlain, D.E., Fuller, R.J., Bunce, R.G.H., Duckworth, J.C. & Shrubb, M. (2000). Changes in abundance of farmland birds in relation to the timing of agricultural intensification in England and Wales. *Journal of Applied Ecology* 37, 771-188.

Champion G.T., May M.J., Bennett S., Brooks D.R., Clark S.J., Daniels R.E., Firbank L.G., Haughton A.J., Hawes C., Heard M.S., Perry J.N., Randle Z., Rossall M., Rothery P., Skellern M.P., Scott R.J., Squire G.R., Thomas M.R. (2003). Crop management and agronomic context of the Farm Scale Evaluations of genetically modified herbicide-tolerant crops. *Philosophical Transactions of the Royal Society London B* 358, 1801-1818.

Chancellor, R.J. (1985). Changes in the weed flora of an arable field cultivated for 20 years. *Journal of Applied Ecology* 22, 491-501.

Costanza, R., d'Arge, R., de Groot, R., Farber, S., Grasso, M., Hannon, B., Naeem, S., Limburg K., Paruelo, J., O'Neill, R.V., Raskin, R., Sutton, P., van den Belt, M. (1997). The value of the world's ecosystem services and natural capital, *Nature* 387, 253–260.

Diaz, S. & Cabido, M. (2001). Vive la difference: plant functional diversity matters to ecosystem processes. *TRENDS* 16, 646-655.

Diaz, S., Symstad, A.J., Chapin III, F.S., Wardle, D.A. & Huenneke, L.F. (2003). Functional diversity revealed by removal experiments. *TRENDS* **18**, 140-146.

Firbank, L.G., Heard, M.S., Woiwod, I.P., Hawes, C., Haughton, A.J., Champion, G.T., Scott, R.J., Hill, M.O., Dewar, A.M., Squire, G.R., May, M.J., Brooks, D.R., Bohan, D.A., Daniels, R.E., Osborne, J.L., Roy, D.B., Black, H.I.J., Rothery, P. & Perry, J.N. (2003). An introduction to the Farm Scale Evaluations of genetically modified herbicide-tolerant crops. *Journal of Applied Ecology* 40, 2-16.

Fuller, R.J., Gregory, R.D., Gibbons, D.W., Marchant, J.H., Wilson, J.D., Baillie, S.R. & Carter, N. (1995). Population declines and range contractions among lowland farmland birds in Britain. *Conservation Biology* 9**,** 1425 – 1441.

Gibbons, D.W., Reid, J.B. & Chapman, R.A. (1993). The new atlas of breeding birds in Britain and Ireland: 1988-1991. London, Poyser.

Gray, A. (1897). The predominance and pertinacity of weeds. *American Journal of Science* 118**,** 161-167.

Gregory, R.D., Noble, D.G. & Custance, J. (2004). The state of play of farmland birds: population trends and conservation status of lowland farmland birds in the United Kingdom. *Ibis* 146 (suppl. 2), 1-13.

Hawes, C., Begg, G., Squire, G.R. & Iannetta, P. (2005). Individuals as the basic accounting unit in studies of ecosystem function: functional diversity in *Capsella* (shepherd's purse). *Oikos* 109, 521-534.

Hawes, C., Haughton, A.J., Bohan, D.A. & Squire, G.R. (2009). Functional approaches for assessing plant and invertebrate abundance patterns in arable systems. *Basic and Applied Ecology* 10, 34-47. doi:10.1016/j.baae.2007.11.007

Hawes, C., Haughton, A.J., Osborne, J.L., Roy, D.B., Clark, S.J., Perry, J.N. et al. (2003). Responses of plant and invertebrate trophic groups to contrasting herbicide regimes in the Farm Scale Evaluations of genetically-modified herbicide-tolerant crops. *Philosophical Transactions of the Royal Society of London B* 358, *1899-1913.*

Heard, M.S., Hawes, C., Champion, G.T., Clark, S.J., Firbank, L.G., Haughton, A.J., Parish, A.M., Perry, J.N., Rothery, P., Scott, R.J., Skellern, M.P., Squire, G.R. & Hill M.I. (2003). Weeds in fields with contrasting conventional and genetically modified herbicide-tolerant crops. I. Effects on abundance and diversity. *Philosophical Transactions of the Royal Society of London Series B-Biological Sciences* 358, 1819-1832.

Hector, A., Dobson, K., Minns, A., Bazeley-White, E. & Lawton, J.H. (2001). Community diversity and invasion resistance: an experimental test in a grassland ecosystem and a review of comparable studies. *Ecological Research* 16, 819-831.

Hewitt, G.M. (1999). Post-glacial re-colonisation of the European Biota. *Biological Journal of the Linnean Society* 68, 87-112.

Holland, J.M., Hutchison, M.A.S., Smith, B. & Aebischer, N.J. (2006). A review of invertebrates and seed-bearing plants as food for farmland birds in Europe. *Annals of Applied Biology* 148**,** 49-71.

Holt, J.S., LeBaron, H.M. (1990). Significance and distribution of herbicide resistance. *Weed Technology* **4**, 141-149.

Iannetta, P.P.M., Begg, G., Hawes, C., Young, M., Russell, J., Squire, G.R. (2007). Variation in *Capsella* (shepherd's purse): an example of intra-specific functional diversity. *Physiologia Plantarum* 129, 542-554.

Imam, A.G. & Allard, R.W. (1965). Population studies of predominantly self-pollinating species. VI. Genetic variability between and within natural populations of wild oats from differing habitats in California *Genetics* **53**, 633-659.

Jones, M.E. (1971). The population genetics of *Arabidopsis thaliana* II. population structure. *Heredity* 27, 51-58.

Kinzig, A.P., Pacala, S.W. & Tilman, D. (eds). (2002). *The Functional Consequences of Biodiversity: Empirical Progress and Theoretical Extensions*. Princeton University Press, Princeton, NJ.

Kleijn, D. & Sutherland, W.J. (2003). How effective are European agri-environment schemes in conserving and promoting biodiversity? *Journal of Applied Ecology* 40, 947-969.

Knops, J.M.H., Tilman, D., Haddad, N.M., Naeem, S., Mitchell, C.E., Haarstad, J., Ritchie, M.E., Howe, K.M., Reich, P.B., Siemann, E. & Groth, J. (1999). Effects of plant species richness on invasion dynamics, disease outbreaks, insect abundances and diversity. *Ecology Letters* **2**, 286-293.

Kogan, M. (1998). Intergrated Pest Management: historical perspectives and contemporary developments. *Annual Review of Entomology* 43, 243–270.

Komeda, Y. (2004). Genetic regulation of time to flower in *Arabidpsis thaliana*. *Annual Review of Plant Biology* 55, 521-535.

Lavorel, S., McIntyre, S., Landsberg, J. & Forbes, T.D.A. (1997). Plant functional classifications: from general groups to specific groups based on response to disturbance. *TREE* **12**, 474-478.

Loreau, M. (1998). Biodiversity and ecosystem functioning: a mechanistic model. *Proceedings of the National Academy of Science* 95, 5632-5626.

Mader, P. Fliebbach, A., Dubois, D., Gunst, L., Fried, P., Niggli, U. (2002). Soil fertility and biodiversity in organic farming. *Science* 296, 1694-1697.

Magurran, A.E. (1988). Ecological diversity and its measurement. Princeton University Press, Princeton, New Jersey.

Marshall, E.J.P. (2004). Agricultural landscapes: field margin habitats and their interaction with crop production. *Journal of Crop Improvement* 12**,** 365-404.

Marshall, E.J.P., Brown, V.K., Boatman, N.D., Lutman, P.J.W., Squire, G.R., Ward, L.K. (2003). The role of weeds in supporting biological diversity within crop fields. *Weed Research* 43, 77-89.

Milton, W.E.J. (1943). The buried viable-seed content of a midland calcareous clay soil. *Empire Journal of Experimental Agriculture* 20, 155-167.

Monteith, J.L. (1977). Climate and the efficiency of crop production in Britain. *Philosophical Transactions of the Royal Society of London B* 281, 277-294.

Moorcroft, D. Whittingham, M.J., Bradbury, R.B. & Wilson, J.D. (2002). The selection of stubble fields by wintering granivorous birds reflects vegetation cover and food abundance. *Journal of Applied Ecology* 39**,** 535-547.

Moreby, S.J. & Southway, S.E. (1999). Influence of autumn applied herbicides on summer and autumn food available to birds in winter wheat fields in southern England. *Agriculture, Ecosystems and Environment* 72, 285-297.

Naeem, S, Wright, J.P. (2003). Disentangling biodiversity effects on ecosystem functioning: deriving solutions to a seemingly insurmountable problem. *Ecology Letters* **6,** 567-579.

Newton, I. (2004). The recent declines of farmland bird populations in Britain: an appraisal of causal factors and conservation actions. *Ibis* 146**,** 579-600.

Owen, M.D.K. (2008). Weed species shifts in glyphosate-resistant crops. *Pest Management Science* 64, 377-387.

Pain, D.J. & Pienkowski, M.W. (1997). *Farming and birds in Europe*. London, Academic Press.

Perkins, A., Whittingham, M.J., Morris, A.J. & Bradbury, R.B. (2002). Use of field margins by foraging yellowhammers *Emberiza citrinella. Agriculture Ecosystems and Environment* 93**,** 413-420.

Potts, G.R. (1991). The environmental and ecological importance of cereal fields. In: Firbank, L.G., Carter, N., Darbyshire, J.F. & Potts, G.R., (eds). *The ecology of temperate cereal fields*. Oxford, Blackwell Scientific Publications. Pp. 3-22.

Powles, S.B. (2008). Evolved glyphosate-resistant weeds around the world: lessons to b e learned. *Pest Management Science* 64, 360-365.

Preston, C.D., Telfer, M.G., Arnold, H.R., Carey, P.D., Cooper, J.M., Dines, T.D., Hill, M.O., Pearman, D.A., Roy, D.B., Smart, S.M. (2002). The changing flora of the UK. Defra, London.

Rands, M.R.W. (1985). Pesticide use on cereals and the survival of Grey Partridge chicks: a field experiment. *Journal of Applied Ecology* 22**,** 49-54.

Roberts, H.A. (1958). Studies on the weeds of vegetable crops. I. Initial effects of cropping on the weed seeds in the soil. *Journal of Ecology* 46, 759-768.

Roberts, H.A. (1962). Studies on the weeds of vegetable crops. II. Effect of six years of cropping on the weed seeds in the soil. *Journal of Ecology* 50, 803-813.

Robinson, R.A. & Sutherland, W.J. (1997). The feeding ecology of seed-eating birds on farmland in winter. In: Donald, P.F. & Aebischer, N.J. (eds). *The ecology and conservation of Corn Buntings* Miliaria calandra. Peterborough, Joint Nature Conservation Committee (UK Nature Conservation, No. 13). Pp. 162-169.

Robinson, R.A. & Sutherland, W.J. (1999). The winter distribution of seed-eating birds: habitat structure, seed density and seasonal depletion *Ecography* 22**,** 447-454.

Robinson, R.A. & Sutherland, W.J. (2002). Post-war changes in arable farming and biodiversity in Great Britain. *Journal of Applied Ecology* 39, 157-176.

Root, R. (1973). Organisation of a plant-arthropod association in simple and diverse habitats: the fauna of collards (*Brassica oleracea*). *Ecological Monographs* 43, 95-124.

Roux, F., Touzet, P., Cuguen, J., Le Corre, V., (2006). How to be early flowering: an evolutionary perspective. *TRENDS in Plant Science* 11, 375-381.

Scherer-Lorenzen, M., Palmborg, C., Prinz, A. & Schulze, E-D (2003). The role of plant diversity and composition for nitrate leaching in grasslands. *Ecology* 84, 1539-1552.

Siemann, E., Tilman, D., Haarstad, J., Ritchie, M. (1998). Experimental tests of the dependence of arthropod diversity on plant diversity. *The American Naturalist* 152 (5), 738-750.

Sheldon, R., Bolton, M., Gillings, S. & Wilson, A. (2004). Conservation management of Lapwing *Vanellus vanellus* on lowland arable farmland in the UK. *Ibis* 146 (suppl. **2**), 41-49.

Sotherton, N.W. (1991). Conservation headlands: a practical combination of intensive cereal farming and conservation. In: Firbank, L.G., Carter, N., Darbyshire, J.F. & Potts, G.R. (eds). *The ecology of temperate cereal fields*. Oxford, Blackwell Scientific Publications. Pp. 373-397.

Sotherton, N.W. & Self, M.J. (2000). Changes in plant and arthropod biodiversity on lowland farmland: an overview. In: Aebischer, N.J., Evans, A.D., Grice, P.V. & Vickery, J.A. (eds). *Ecology and conservation of lowland farmland birds*. Tring, British Ornithologists Union. Pp. 26-35.

Squire, G.R., Brooks, D.R., Bohan, D.A., Champion, G.T., Daniels, R.E., Haughton, A.J., Hawes, C., Heard, M.S., Hill, M.O., May, M.J., Osborne, J.L., Perry, J.N., Roy, D.B., Woiwod, I.P., Firbank, L.G. (2003). On the rationale and interpretation of the farm-scale evaluations of genetically-modified herbicide-tolerant crops. *Philosophical Transactions of the Royal Society of London B* **358**, 1779-1800.

Squire, G.R., Hawes, C., Begg, G.S., Young, M. (2009). Cumulative impact of GM herbicide tolerant cropping on arable plants assessed through species-based and functional taxonomies. Environmental Science and Pollution Research 16, 85-94

Squire, G.R., Rodger, S., Wright, G. (2000). Community-scale seedbank responses to less intense rotation and reduced herbicide input. *Annals of Applied Biology* 136, 47-57.

Storkey, J. (2006). A functional group approach to the management of UK arable weeds to support biological diversity. *Weed Research* 46, *513-522.*

Swash, A.R.H., Grice, P.V. & Smallshire, D. (2000). The contribution of the UKBiodiversity Action Plan and agri-environment schemes to the conservation of farmland birds in England. In: Aebischer, N.J., Evans, A.D., Grice, P.V. & Vickery, J.A. (eds). *Ecology and conservation of lowland farmland birds*. Tring, British Ornithologists Union. Pp. 36-42.

Taylor, R.L., Maxwell, B.D. & Boik, R.J. (2006). Indirect effects of herbicides on bird food resources and beneficial arthropods. *Agriculture Ecosystems and Environment* 116, 157-164.

Thompson, D.B.A. & Barnard, C.J. (1983). Anti-predator responses in mixed species associations of Lapwings, Golden Plovers and Black-headed Gulls. *Animal Behaviour* 31, 585-593.

Tilman, D., Knops, J., Wedin, D., Reich, P., Ritchie, M. & Siemann, E. (1997). The influence of functional diversity and composition on ecosystem processes. *Science* 277, 1300-1302.

Tucker G.M. (1992). Effects of agricultural practices on field use by invertebrate-feeding birds in winter. *Journal of Applied Ecology* 29, 779-790.

Tucker, G.M. & Heath, M.F. (1994). *Birds in Europe: their conservation status*. Cambridge, BirdLife International (BirdLife Conservation Sries no. 3).

Whittingham, M.J. & Evans, K.L. (2004). The effects of habitat structure on predation risk of birds in agricultural landscapes. *Ibis* 146 (suppl. 2), 210-220.

Wilson, J.D., Taylor, R. & Muirhead, L.B. (1996). Field use by farmland birds in winter: an analysis of field type preferences using resampling methods. *Bird Study* 43, 320-332.

Yan, L., Loukoianov, A., Tranquilli , G., Helguera, M., Fahima, T., Dubcovsky, J. (2003). Positional cloning of a wheat vernalisation gene *VRN1*. *Proceedings of the National Academy of Science USA* 100, 6263-6268.

Young, J.E.B., Griffin, M.J., Alford, D.V., Ogilvy, S.E. (2001). Reducing agrochemical use on the arable farm - The Talisman and Scarab projects. Defra, London.

In: Weeds: Management, Economic Impacts and Biology
Editor: Rudolph V. Kingely
ISBN: 978-1-60741-010-2

Short Communications

BIOPHOTONS AS A NOVEL INDICATOR OF HERBICIDAL SELECTIVITY IN PLANTS

Kmihiko Kato
National Agricultural Research Center, Tsukuba, Japan

1. WHAT ARE BIOPHOTONS?

All living organisms, including fungi, plants, and mammals, spontaneously produce ultraweak photon emissions that originate from cellular biochemical reactions. This luminescence is commonly referred to as biophoton. The emission intensity of biophotons is extremely weak at 10^0–10^3 photons·$(s^{-1}·cm)^{-2}$, and they are considered to be associated with the oxidative metabolic responses essential for life activities, unlike the bioluminescence observed in fireflies, which employs a luciferin-luciferase system (Fig.1). The detection of biophotons has been rendered possible by the development of photomultiplier tubes, and by using such a device, Coli (1955) was the first to observe biophotons in seedlings of plants such as wheat and lentils. Since then, this ultraweak photon emission has been reported by many researchers at the organism, tissue, and cellular levels (Pop *et al.,* 1992; Beloussov *et al.,* 2000). It has been variously suggested that reactive oxygen species are the driving force of biophoton generation because biophotons disappear if the oxygen supply is stopped (Abeles, 1986). Majority of biophotons, if not all, are believed to be emitted in the process of oxidation of substances, including unsaturated fatty acids, amino acids, and polyphenols. These substances are peroxidized and excited by reactive oxygen species (ROS) or enzymes such as NADPH oxidase, peroxidase, lipoxygenase, and cytochrome P450. Excited carbonyl compounds generated through the process are considered to be important emitting molecules (Abeles, 1986). In some cases, this excited carbonyl compound is produced as a result of free radical recombination reactions (Radotic *et al.*, 1998). However, because the quantum efficiency of excited carbonyl is relatively low, the biophotons detected from the living body, to some extent, appear to originate in fluorescent substances in cells, to which excess energy of exited carbonyl is transferred (Watanabe and Inaba, 1991). In plants, chlorophyll is a good example of this energy transfer. In green seedlings containing large amounts of chlorophyll, biophoton emission was observed only in the range of red light whereas its original spectrum of biophoton emission was between yellow to orange (Slavinski *et al.,* 1981). In addition, among ROS, only singlet oxygen species emit light. However, except in special cases, the concentration of singlet oxygen in the cell is not sufficiently high to produce biophotons, which can be detected with a photon counter (Watanabe and Inaba, 1991). Therefore, the detailed mechanism of biophoton emission in the living body has not been fully clarified.

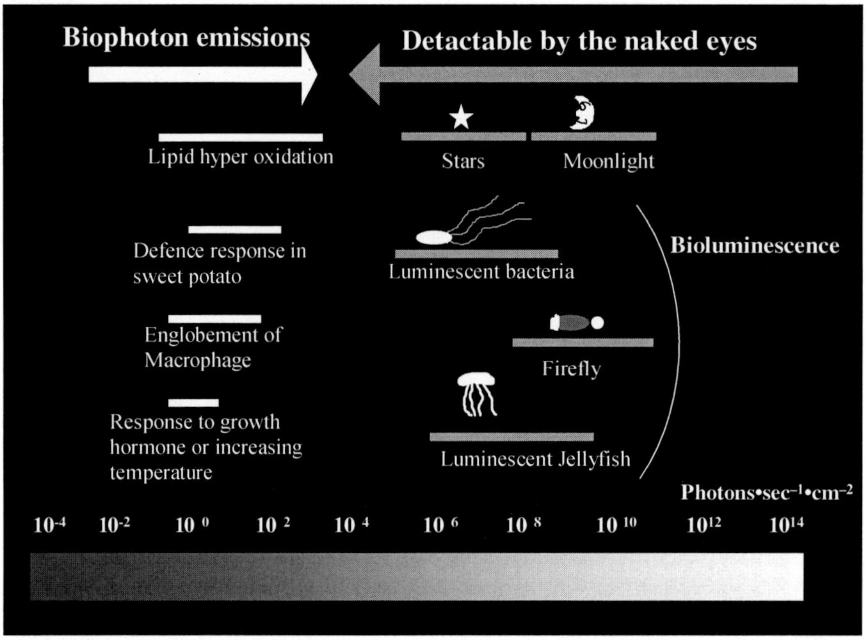

Figure 1. Difference in the emission intensities of bioluminescence and biophoton emissions.

Biophotons can be detected easily, nondestructively, and in real time by using a photon counter equipped with highly sensitive photomultiplier tubes. Therefore, biophotons have gained considerable attention as a novel tool for identifying the condition of organisms. Biophoton research is particularly advanced in the medical field, where the device has been experimentally applied to tumor detection (Musumeci *et al.,* 1992), brain monitoring (Kobayashi *et al.,* 1999) and renal failure diagnosis (Agatsuma *et al.,* 1992).

2. Biophoton Emission from Plants in Response to Stress.

In plants, the most important feature of biophotons is to produce responses to environmental stimuli, in which biophoton increments are primarily observed in plants that acclimatize to these stimuli. Environmental stimuli include biological (e.g., pathogens, insects, and wounds) and non-biological (e.g., temperature, drought, and salt concentration). Biophoton emission occurs in response to destruction of cells and as a result of controlled biochemical reactions in cells. Many studies have attempted to detect stress responses in plants using biophotons. Some of these studies have been cited below.

2-1. Biophoton Emission in Response to Biological Stimuli

It is well known that pathogenic infections induce a significant increase in biophoton emission in plants. On the basis of the disease-resistance reactions of plants, biophotons are classified into 2 types: the relatively weak emissions observed during the early stages of resistance reaction (Iyozumi *et al.,* 2005) and strong emission intensities in cells exhibiting programmed cell death (PCD) during the middle stages of resistance reaction to localize pathogen from healthy cells (Bennett *et al.,* 2005). In sweet potatoes undergoing PCD as a result of inoculation with *Fusarium oxysporum*, it was reported that the wavelength composition of photon emission considerably shifted toward a shorter wavelength as compared with that of untreated samples, indicating that this was a luminous phenomenon qualitatively different from the one observed under normal conditions (Iyozumi *et al.,* 2002). In addition, in the early stage of resistance reaction, biophoton emission has been proven to occur via the signaling cascade occurring in resistance reactions (Kageyama *et al.,* 2006). Taking into consideration the first report on the direct link between resistance reaction and biophotons, the latter report presents a very important finding. Because of this direct linkage, biophotons were enhanced during the early resistance reaction in plants pretreated with plant activators that can enhance the resistance reaction to pathogens (Iyozumi *et al.,* 2005). Thus, the 2 types of biophoton generation found in disease-resistance reactions are clearly based on controlled biochemical oxidative reactions in cells.

Plant hormones are another example of biological stimuli. Exogenously applied plant hormones cause a change in the biophoton emission from plants. When plants were treated with gibberellins, auxins, salicylic acid, and jasmonic acid, there was a mild to strong increase in the biophoton emission (Kai *et al.*, 1995; Iyozumi *et al.,* 2002). On the other hand, when abscisic acid (ABA) was applied to plants, the biophoton emission was strongly inhibited. In this case, it was suggested that the stress response induced by ABA either inhibited the generation or enhanced the scavenging of ROS. It has also been reported that when auxins were applied to sections of the sweet potato tuberous root, the emission wavelength shifted toward a shorter wavelength as compared to that of untreated samples (Iyozumi *et al.,* 2002). Plant hormone-related biophotons appear to reflect the change in controlled biochemical oxidative reactions.

2-2. Biophoton Emission Response to Non-Biological Stimuli

Biophoton emission from plants has been reported to dramatically increase under temperatures that are fatally high for plants (Hauvaux, 2003). This is a good example of biophoton emission in response to cell destruction, in which uncontrolled oxidative reactions occur. Further, an increase in biophoton emission was observed in the process of acclimation of plants to high temperatures (but not fatally so) (Iyozumi et al., 2002). In contrast, since a transient increase in biophotons is also caused by a sudden decrease in temperature, it has been proposed that this phenomenon can be applied for testing the resistance of plants to frost damage (Agaverdiyev, 1965).

Plants also exhibit biophoton generation responding to salt and drought stresses. When azuki bean sprouts were treated with NaCl solution, biophoton emission weakened until the concentration reached approximately 1 M, but it increased at 4.5 M. (Ohya, 2000). This phenomenon is considered to be derived from the generation of reactive oxygen as a result of electron leakage from mitochondria caused by high-concentration ions. A mild increase in biophoton emission also occurred under dry conditions, and an extremely strong increase in biophoton intensity was observed when the sprouts that had been dried over many hours were permitted to absorb water again (Ohya, 2002). This burst of biophoton emission probably resulted from a dramatic promotion of oxidative degradation of cellular elements, which was caused by oxidases leaked from the dry, damaged cells when the cells absorbed water. Thus, under conditions of severe stress that are scarcely conducive for the survival of plants, the increment in biophoton emission is a result of the destructive oxidative process of cells.

3. Possibility for the Identification of Herbicide-Resistant Weed Biotypes By Detecting Biophoton Emission

As described above, biophotons have diverse potential as a source of comprehensive biological information that might help us understand the physiological condition of plants. Biophotons are considered to be particularly useful for examining the stress responses of plants. Herbicide treatment is a severe stress factor for plants. Although studies dealing with the effects of herbicide treatment on biophoton emission are limited, it has been demonstrated that biophotons can be used as a novel indicator of resistance biotypes in weeds.

3-1. Herbicide Resistance in Weeds.

In the 1950s, a herbicide-resistance biotype in weeds was first reported to be resistant to the herbicide 2,4-dichlorophenoxyacetic acid (2,4-D) (Hilton, 1957). Since then, the number of weed biotypes resistant to various herbicides has increased dramatically and such biotypes have been reported worldwide (Heap, 2008). Currently, there are more than 320 biotypes in 185 weed species that have evolved resistance to compounds in all the major groups of herbicides.

There are 2 primary mechanisms of herbicide resistance in weeds (Dekker and Duke, 1955; Yuan *et al.,* 2006). One is attained by avoiding herbicidal effects through the mutation or excessive expression of herbicide target molecules. This is termed target-site resistance. Many weed populations worldwide have developed target-site resistance. Weeds with target-site resistance to one herbicide group are usually susceptible to herbicides belonging to a different group. Examples include the reduced sensitivity to target enzymes of acetyl-CoA carboxylase (ACCase), acetolactate synthase (ALS), and 5-enolpyruvylshikimate 3-phosphate synthase (EPSPS).

On the other hand, plants have another resistance mechanism by which they limit the amount of herbicide reaching the target site. For example, herbicides are metabolized before they can reach the target sites. This is termed non-target site resistance. Non-target site resistance endows resistance to one or few similar herbicides. Many plant detoxifying proteins might be involved in non-target site herbicide resistance. Thus far, it is known that non-target site herbicide resistance has been well established in only 4 gene families: P450s,

GSTs, glycosyltransferases, and ABC transporters. Of these, only P450s participate in oxidative reactions, catalyzing oxygen- and NADPH-dependent mono-oxygenation.

3-2. Biophoton Emission from Herbicide-Resistant Weeds.

Paraquat is a typical photosystem I inhibitor. It acts on the photosynthetic membrane system in photosystem I. Paraquat causes tissue damage by generating ROS through reduction of molecular oxygen. Intense biophoton emission was observed from leaves of tobacco treated with paraquat (Fig.2) (Kato *et al.,* 2003). Furthermore, it was confirmed that biophoton emission from leaves treated with paraquat decreased when antioxidants such as catechin were additionally applied (Kato *et al.,* 2003). Therefore, it was suggested that the biophoton emission from paraquat treatment was strongly related to the ROS generated in cells.

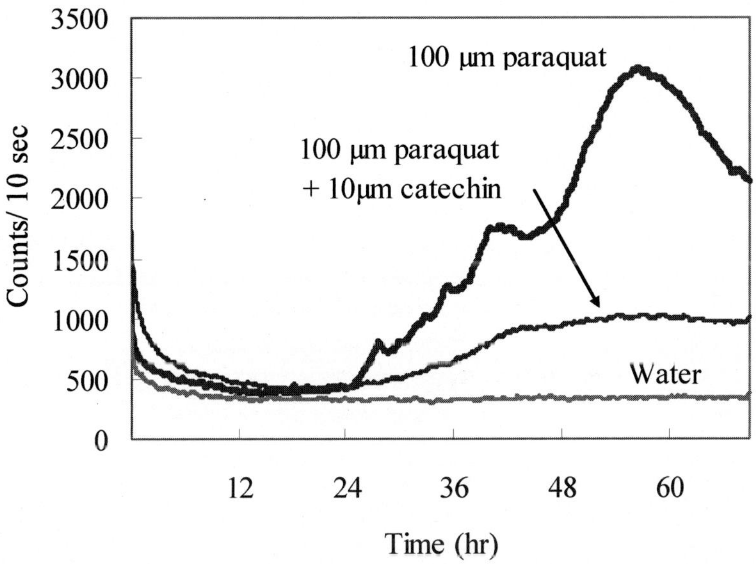

Figure 2. Effectiveness of catechin on biophoton emissions from leaves of tobacco treated with paraquat (Kato et al., 2003).

Hideg and Inaba (1991) demonstrated that the leaves of paraquat-resistant tobacco treated with paraquat emitted weaker photons as compared with tobacco plants sensitive to paraquat. The superoxide dismutase (SOD) activity of the resistant plant was 3 to 6 times greater than that of sensitive plants. Apparently, higher SOD activity in resistant tobacco contributes to the elimination of the ROS induced by paraquat, resulting in weaker photon emission. Thus, biophoton emission can be used to identify weeds resistant to paraquat.

In the above example, biophoton emission decreases in biotypes that are resistant to herbicides; however, studies with opposing results have also been reported. Sulfonylurea (SU) is one of the most potent ALS-inhibiting herbicides, and biotypes resistant to SU are detected in many weed species. The resistance mechanisms of weeds to SU are considered target-site resistance almost without exception. Visual effects of SU herbicides on weeds develop slowly because SU targets the enzyme that catalyzes the biosynthesis of branched-chain amino acids and inhibits this enzyme (Stidham, 1991); hence, visual determination of the effects of SU takes a long time. Identification of resistant biotypes is essential for effective weed control. Therefore, various detection methods such as bioassays and *in vivo* and *in vitro* assays have been developed (Gerwick *et al*., 1993; Simpson *et al*., 1995; Lovell et al., 1996; Boutsalis, 2001). Of these methods, biophoton emission assay appears to be one of the most useful methods for identifying resistant weed biotypes. In an experiment employing 12 lines of *Scirpus juncoides*, the intensity of biophoton emission differed between the SU-resistant and SU-susceptible biotypes, i.e., the biophoton emission intensity after SU (bensulfuron-methyl) treatment was higher in the former than in the latter (Fig.3) (Inagaki *et al.,* 2008). This increment was recorded, irrespective of the differences in the mutation sites of the ALS gene (target site of SU). Furthermore, this difference in biophoton emission between the resistant and susceptible biotypes was observed from the vegetative growth stage to the flowering stage (Inagaki *et al.,* 2008). Therefore, it is suggested that biophoton emission could be an indicator of SU-resistant biotypes in various weeds. In fact, the difference in biophoton emission observed in the resistant and susceptible biotypes in *S. juncoides* has also been confirmed in another paddy weed (Inagaki *et al.,* unpublished).

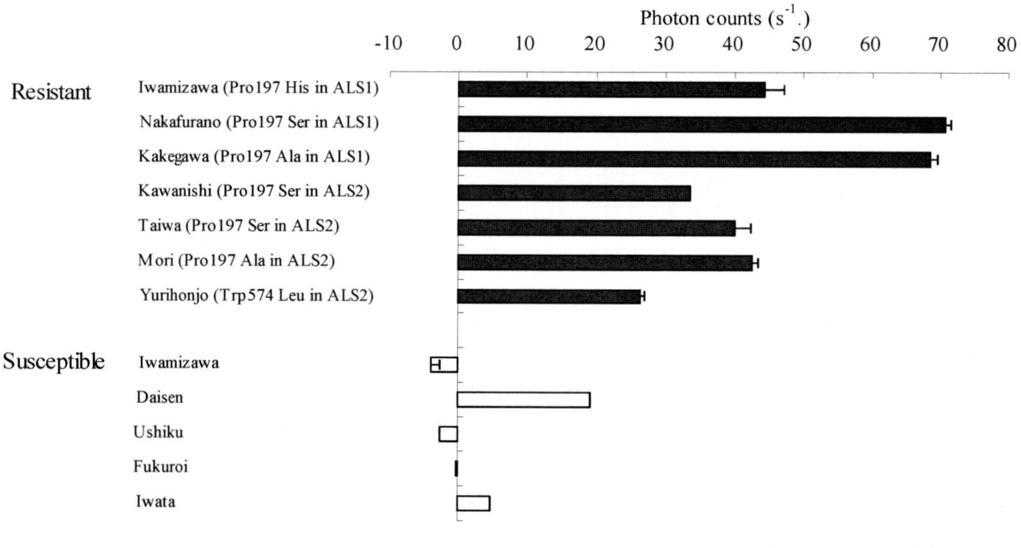

Figure 3. Increment of biophotons in culms of sulfonylurea-resistant and sulfonylurea-susceptible biotypes in Scirpus juncoides treated with sulfonylurea. Bars indicate standard deviations (±SD). (Inagaki et al., 2008)

Although weed species developed resistance to SU herbicides due to target site resistance, rice and barnyard grass are innately resistant to SU herbicides because of non-target site resistance. Intense biophoton emission is observed from rice and barnyard grass treated with SU herbicides (Fig.4) (Inagaki *et al.,* 2007). The SU resistance of rice and barnyard grass occurs due to detoxification metabolism. In particular, P450 plays a central role in this resistance because SU resistance of rice was repressed by the inhibition of P450 activity (Shirakura *et al.,* 1995). Biophoton emissions from rice and barnyard grass treated with SU herbicide were suppressed upon pretreatment with P450 inhibitors, piperonyl butoxide, and malathion (Inagaki *et al.,* 2007). It is therefore suggested that oxidative metabolism by P450 is strongly associated with the increase in biophoton emission caused by SU herbicides in these plants. This biophoton emission during the process of mono-oxygenation with P450 obviously occurs under a controlled system in plants.

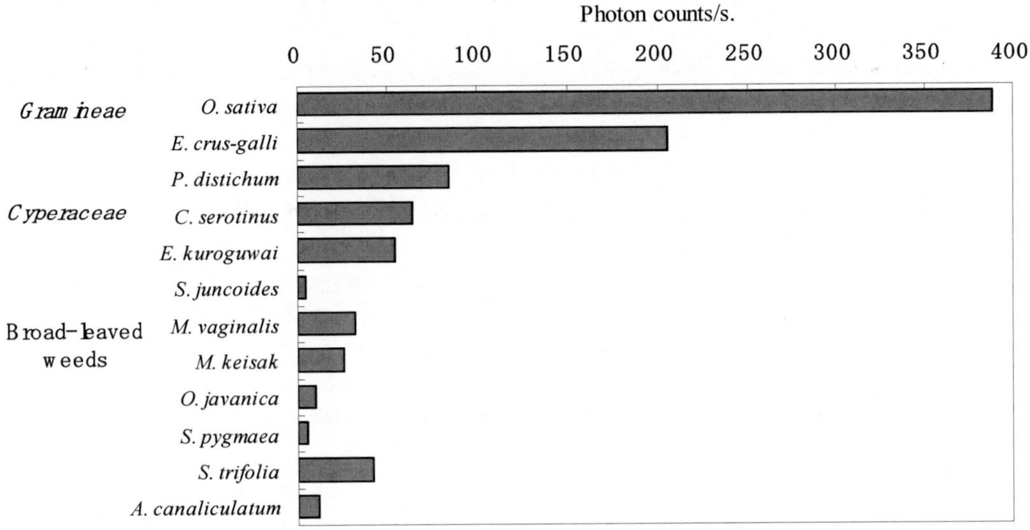

Figure 4. Increases in the biophoton emissions from leaves of rice and paddy weed species. (Inagaki et al., 2007)

In case of non-target site resistance to SU, P450 clearly plays a major role in biophoton generation, but the mechanism of target site resistance to SU has not yet been elucidated. We hypothesize that SU, which cannot bind to ALS due to ALS mutations, would also be detoxified with P450, resulting in the increase in biophoton emission. This hypothesis is supported by the experimental result in which biophoton emission from the resistant paddy weed biotypes treated with SU was suppressed by P450 inhibitors (Inagaki *et al.,* unpublished). We consider that if P450 is involved in herbicide resistance, regardless of non-target-site or target-site type of mechanism or whether or not P450 plays a major role in herbicide resistance, herbicide-resistant biotypes can be identified by evaluating biophoton emission.

3-3. Comprehensive Importance of Biophoton Emission in Diagnosis of Herbicide-Resistant Weeds

As described above, biophoton can be potentially used in a novel diagnosis system as an indicator of weeds resistant to herbicide. However, because oxidative biochemical reactions are necessary for biophoton generation, there are obvious limitations.

ROS production in cells due to the action of paraquat is directly linked to biophoton generation. This is a typical example of biophoton generation directly resulting from the effect of the herbicide. Similar to paraquat, there are several types of herbicides that can induce increases in ROS in cells. Herbicides whose sites of action are inhibition of photosynthesis or respiration would belong to this category. When plants are sprayed with such herbicides, the herbicide-sensitive plants are expected to show increases in biophoton emissions; however, biophoton emissions from herbicide-resistant plants would decrease because of their higher capability of quenching ROS.

Biophoton emission can be also used to detect herbicide resistance based on P450. This is a good example of biophoton emission caused by controlled oxidative reactions in plants. As explained before, the resistant biotypes mediated by P450 can be identified using biophoton emission. Since P450 plays a major role in the detoxification of resistant weeds, these biotypes have been found in many important weed species, such as blackgrass, chickweed, rigid ryegrass, and wild mustard (Yuan *et al.,* 2006). Further, we would like to re-emphasize that resistant weeds can be identified using biophoton emission, regardless of the role played by P450 in herbicide resistance.

Thus, biophoton emission can be used in the identification of herbicide-resistant weeds, provided oxidative reactions in cells are induced by herbicide application. In addition, except for P450, if the lipoxygenases in resistant weeds are stimulated by herbicide application, the biophoton identification method may be applied for these weeds. However, as mentioned before, there is a limitation for the application of biophoton to indicate herbicide resistance. It might be difficult to detect resistance in resistant biotypes whose herbicide-resistance mechanisms are not entirely related to oxidative reactions. For example, it is expected that identification of resistant biotypes would be difficult in non-target site resistance mechanisms, which are mediated by ABC transporters, GSTs, and glycosyltransferases where P450s are not involved in the mechanism. In fact, the biophoton emission from a resistant plant treated with sulfometuron-methyl, which is not metabolized by P450 in that plant, would scarcely increase (Inagaki *et al.* unpublished).

At present, our knowledge on the application of biophoton is limited to the diagnosis of herbicide-resistant weed species. Further studies with more combinations of herbicides and their resistant weed species are required. Furthermore, it is important to clarify the detailed mechanism of biophoton emission. The elucidation of the biophoton mechanisms and further development of practical detection techniques are the key issues to be resolved.

REFERENCES

Colli, L, Facchini, U, Guidotti, G, Lonati, RD, Orsenigo, M, Sommariva O. 1955. Further measurements on the bioluminescence of seedlings. Experientia 11:479–81.

Popp, FA, Li, KH, Gu, Q, Editors. 1992. Recent advances in biophoton research and its applications. World Scientific, Singapore.

Beloussov, L, Popp, FA, Voeikov, V, Wijk, RV, Editors. 2000. Biophotonics and Coherent Systems, Moscow University Press, Moscow.

Abeles, FB. 1986. Plant chemiluminescence. Annu. Rev. Plant Physiol. 37: 49–72.

Radotic, K, Radenovic, C, Jeremic, M. 1998. Spontaneous ultraweak bioluminescence in plants: origin, mechanisms and properties. Gen. Physiol. Biophys. 17: 289–308.

Suzuki, S, Usa, M, Nagoshi, T, Kobayashi, M, Watanabe, N, Watanabe, H, Inaba, H. 1991. Two-dimensional imaging and counting of ultraweak emission patterns from injured plant seedlings. J. Photochem Photobiol B Biol. 9: 211–217.

Slawinski, J, Grabikowski, E, Ciesla, L. 1981. Spectral distribution of ultraweak luminescence from germinating plants. J. Luminescence 24/25: 791–794.

Musmeci, F. 1992 Recent advances in biophoton research and its applications. 302–324, World Scientific, Singapore.

Kobayashi, M, Takeda, M, Ito, K, Kato, H, Inaba, H. 1999. Two-dimensional photon counting imaging and spatiotemporal characterization of ultraweak photon emission from a rat's brain in vivo. J. Neurosci. Methods 93: 163–168

Agatsuma, S, Nagoshi, T, Kobatashi, M, Usa, M, Watanabe, H, Sekino, H, and Inaba, H. 1992. Hydroxyl radical-induced characteristics chemiluminescent spectra from plasma of hemodialysis patients. Clin. Chem. 38: 48–55.

Iyozumi, H, Kato K, Kageyama, C, Inagaki, H, Yamaguchi, A, Furuse, K, Baba, K, Tsuchiya, H. 2005. Plant defense activators potentiate the generation of elicitor-responsive photon emission in rice. Physiol. Mol. Plant Pathol. 66: 68–74.

Bennett, M, Mehta, M, Grant, M. 2005. Biophoton imaging: a nondestructive method for assaying R gene responses. Mol. Plant-Microbe Interact. 18: 95–102.

Iyozumi, H, Kato, T, Makino, T. 2002. Spectral shift of ultraweak photon emission from sweet potato during a defense response. Photochem. Photobiol. 75: 322–325.

Kageyama, C, Kato K, Iyozumia, H, Inagaki, H, Yamaguchi A, Furuse, K, Baba, K. 2006. Photon emissions from rice cells elicited by N-acetylchitooligosaccharide are generated through phospholipid signaling in close association with the production of reactive oxygen species. Plant Physiol. Biochem. 44 : 901–909.

Kai, S, Ohya, T, Moriya, K, Fujimoto, T. 1995. Growth control and biophoton radiation by plant hormones in red bean. Jpn. J. Appl. Phys. 34: 6530–6538.

Hauvaux, M. 2003. Spontaneous and thermoinduced photon emission: new methods to detect and quantify oxidative stress in plants. Trends in Plant Science 8: 409–413.

Agaverdiyev, AS, Doskoch, YE, Tarusov, BN. 1965. Effect of low temperatures on the ultraweak luminescence of plants. *Biofizika* 10: 832–836.

Ohya, T, Kurashige, H, Okabe, H, Kai, S. 2000. Early detection of salt stress damage by biophotons in red bean seedlings. Jpn. J. Appl. Phys. 39: 3696–3700.

Ohya, T, Yoshida, S, Kawabata, R, Okabe, H, Kai, S. 2002. Biophoton emission due to drought injury in red beans: possibility of early detection of drought injury. Jpn. J. Appl. Phys. 41: 4766–4771.

Hilton, HW. 1957. Herbicide tolerant strain of weeds. Hawaiian Sugar Planters Association Annual Reports. pp. 69.

Heap I. 2008. International survey of herbicide-resistant weeds. Available at http://www.weedscience.org/. Accessed August 6, 2008.

Dekker, J. and Duke, O. S. 1995. Herbicide-resistant field crops. Adv. Agron. 54: 69–116.

Yuan, JS, Tranel, PJ, Stewart, CN. Jr. 2006. Non-target-site herbicide resistance: a family business. Trends Plant Sci. 12: 6–13.

Kato, K, Makino, T, Matsuura, H. 2003. Japanese Patent 2004-233162.

Hideg, É. and Inaba, H. 1991. Dark adapted leaves of paraquat-resistant tobacco plants emit less ultraweak light than susceptible ones. Biochem. Biophys. Res. Commun. 178: 438–443.

Stidham, MS. 1991. Herbicides that inhibit acetohydroxy acid synthase. Weed Sci. 39, 428–434.

Gerwick, BC, Mireles, LC, Eilers, RJ. 1993. Rapid diagnosis of ALS/AHAS-resistant weeds, Weed Technol. 7: 519–524.

Simpson, DM, Stoller, EW, Wax, LM. 1995. An in vivo acetolactate synthase assay, Weed Technol. 9: 17–22.

Lovell, ST, Wax, LM, Simpson, DM, McGlamery, M. 1996. Using the in vivo acetolactate synthase (ALS) assay for identifying herbicide-resistant weeds, Weed Technol. 10: 936–942.

Boutsalis, P. 2001. Syngenta quick-test: A rapid whole-plant test for herbicide resistance. Weed Technol. 15: 257–263.

Inagaki, H, Ishida, Y, Uchino, A, Kato, K, Kageyama, C, Iyozumi, H, Nukui, H. 2008. Difference in ultraweak photon emissions between sulfonylurea-resistant and sulfonylurea-susceptible biotypes of *Scirpus juncoides* following the application of a sulfonylurea herbicide. Weed Biol. Management. 8: 78–84.

Inagaki, H, Imaizumi, T, Wang, GX, Tominaga, T, Kato, K, Iyozumi, H, Nukui, H. 2007. Spontaneous ultraweak photon emission from rice (*Oryza sativa* L.) and paddy weeds treated with a sulfonylurea herbicide. Pestic. Biochem. Physiol. 89: 158–162.

Shirakura, S, Ito, K, Aizawa, H. 1995. Effect of Cytochrome P-450 Monooxygenase Inhibitors on the Rice Tolerance of Azimsulfuron and Bensulfuron-methyl. Weed Res. 40: 218–220.

In: Weeds: Management, Economic Impacts and Biology
Editor: Rudolph V. Kingely
ISBN: 978-1-60741-010-2

Short Communications

WEED CONTROL IN MAIZE (*ZEA MAYS* L.) WITH ATRAZINE AND ITS SOIL RESIDUAL ACTIVITY ON THE GROWTH AND YIELD OF AMARANTH (*AMARANTHUS CRUENTUS* (L.) SAUER) IN SOUTHWESTERN NIGERIA

R.D. Aladesanwa* and S.A. Adejoro

*Department of Crop, Soil and Pest Management, The Federal University of Technology, P.M.B. 704, Akure, Ondo State, Nigeria

ABSTRACT

Atrazine, despite being banned in most European countries, is still widely used in the United States and is registered in more than 70 countries worldwide. It is still by far the most widely used symmetrical triazine herbicide for weed control in maize in southwestern Nigeria, following which vegetable crops may be sown to 'catch' remaining moisture under rainfed multiple cropping. A randomized complete block field experiment was conducted over 2 years to evaluate the influence of residual treatments of atrazine at 1.0, 2.0, 3.0, and 4.0 kg a.i. / ha from maize cropping, including a weedy and weed-free controls on the growth and yield of amaranth at the Teaching and Research Farm of the Federal University of Technology, Akure located in the rainforest vegetation zone of Nigeria. Results indicated that atrazine at 3.0 and 4.0 kg a.i. / ha provided effective weed control and significantly ($p<0.05$) increased maize grain yield over the weedy control at a similar level with the weed-free control, but proved completely ineffective at 1.0 and 2.0 kg a.i. $ha^{-1.}$ Results further showed that atrazine at all the rates tested did not exert any adverse residual effects on the subsequent growth and yield of amaranth in both years. Consequently, it is concluded that amaranth may be sown to succeed maize in rotation without suffering a concomitant reduction in crop growth and yield under rainfed multiple cropping.

Key words: Soil residual activity, maize and *Amaranthus cruentus*

INTRODUCTION

Atrazine is a prominent triazine and one of the most widely used herbicides in world agriculture (Alvi *et al.*, 2003). In spite of being banned in most European countries, it is still widely used in the United States and is registered in more than 70 countries worldwide (Kauffmann *et al.*, 2000). It is still by far the most widely used symmetrical triazine herbicide for weed control in maize in southwestern Nigeria, following which vegetable crops may be sown to 'catch' remaining moisture under rainfed multiple cropping. It shows good selectivity in maize and sorghum and is highly active in the soil against a wide range of broadleaved weeds and annual grasses (Klingman and Ashton, 1975). The appeal for its continued use in this part of the world can be attributed to its ready availability, cheapness and effectiveness over the years (Aladesanwa, 2005).

Because of its high soil activity, the length of time that atrazine remains active in the soil is of particular importance since this could relate to phytotoxic after-effects that may prove injurious to succeeding crops or plantings. For example, it has been established that atrazine at 1.5 kg a.i. / ha may damage soybeans grown in the year following maize (McFarlane, 1977). In contrast, Akinyemiju *et al.* (1986) reported that atrazine at the foregoing dosage rate was no longer phytotoxic to tomato seedlings under field conditions 8 weeks after application in the humid tropical environment.

At the moment information on the soil residual activity of atrazine on the growth of most vegetables in the tropics is lacking. Since any of these vegetables may be sown after maize crop in which atrazine has been used for selective weed control, it then becomes necessary to gain an insight into the soil residual activity of atrazine on their growth and development in order to prevent crop injury.

Consequently, the main objective of the present study was to evaluate the influence of atrazine residue from maize cropping on the growth and yield of amaranth, an important tropical leafy vegetable crop rich in vitamin A, calcium and potassium (Kochhar, 1986). To achieve the foregoing objective, the effects of atrazine on weed growth, soil nutrient content at maize harvest and maize grain yield were also monitored in the different treatments prior to the sowing of the vegetable, because such an holistic approach would permit an unbiased estimate of residual treatment effects of atrazine on the subsequent growth of amaranth under rainfed multiple cropping.

MATERIALS AND METHODS

The study was conducted at the Teaching and Research Farm of the Federal University of Technology, Akure (7°16′N, 5°12′E) located in the rain forest vegetation zone of Nigeria during the late (September-December) and early (April-July) cropping seasons of 2002 and 2003, respectively. The average annual rainfall is about 1300mm with a mean temperature of 27°C and the climate is of the subhumid type. The soil at the experimental site in both years was a sandy clay loam. Chemical analyses of the top soil at 5cm depth before sowing commenced are given in Table 1.

A preliminary field trial involving application of atrazine in maize at 1.0, 2.0, 3.0 and 4.0 kg a.i. / ha including a weed-free control, where weeding occurred once per week till harvest at 12 WAP and a weedy control, where no weeding occurred during the experimental period, was laid out in a randomized complete block design (RCBD) with four replications per treatment. Two seeds of Downy mildew resistant (DMR) maize variety were sown per stand but later thinned to one seedling per stand at 2 weeks after planting (WAP). Planting was done on plots measuring 6 × 4 m at a spacing of 75 × 25 cm with 1 m alleyway between plots. Atrazine (480F), a flowable formulation, was applied early postemergence to both weed and crop at the foregoing rates with a knapsack sprayer fitted with Polijet nozzles calibrated to deliver 250 l / ha of the spray solution at a pressure of 2.5 kg / cm^2.

Table 1. Nutrient status and soil pH of experimental sites before sowing maize

Year	pH	Organic matter (%)	N (%)	P (mg / kg)	K (cMol / kg soil)	Ca (cMol / kg soil)	Mg (cMol / kg soil)
2002	6.4	3.2	0.3	15.4	0.3	1.7	0.6
2003	6.3	3.3	0.3	16.0	0.3	2.0	0.6

The main weed population in the experimental area in both years comprised both broadleaved and grass species, prominent among which were *Aspilia africana, Euphorbia heterophylla, Spigelia anthelmia, Cynodon dactylon, Eleusine indica, Panicum maximum* and *Pasperlum orbiculare*.

Weed assessment was conducted twice at 4 weeks after treatment (WAT) and at harvest to determine the weed spectrum, density and weight of total weed species using three fixed 50 × 50 cm quadrats along a diagonal in each plot from which weed samples were collected and analyzed for the foregoing parameters. Collected weed samples were bulked and subsequently weighed. At harvest, all the plots were clean-weeded regardless of previous weeding treatments, and the weeds weighed, thus allowing a comparison to be made of the total weed growth from the time of sowing till harvest for the different treatments. All treatments were weed free at sowing. Soil samples were also collected at harvest at 0-15 cm depth in all the plots to determine the soil pH and nutrient status using standard methods (AOAC, 1990).

Yield was assessed from ten randomly selected plants per plot. The harvested cobs were shelled, weighed and grain weight adjusted to 13% moisture. Following maize harvest at 12 WAP, seeds of amaranth were planted on raised beds prepared on demarcated plots from the preliminary maize trial at a spacing of 30 × 10 cm^2.

Weeds were manually controlled in all plots once at 2 and 5 WAP with a hand hoe. The experiment was terminated each year at 8 WAP when the leaves of the amaranth were fully matured for consumption. A sample area of 1 m^2 was harvested from each plot for the determination of growth and yield components. Edible as well as marketable yield was estimated based on leaf fresh weight plus stem weight. Total biomass was determined based on leaf fresh weight plus stem weight and root weight. Leaf area was determined using the graph paper method.

Data collected from the experiments in both years were submitted to an analysis of variance while treatment means were compared using the standard error of mean differences. Simple linear correlation and regression analysis between increasing dose of atrazine (X) and soil pH, soil nutrient content, growth or yield parameters (Y) of the amaranth was performed with a scientific calculator (Casio *fx*-7400G PLUS POWER GRAPHIC Model).

RESULTS

Effects of Atrazine on Weed Growth and Maize Grain Yield

Data collected during the experiments are shown in the tables. There were significant ($p<0.05$) differences amongst treatments in weed weight recorded at 4 WAT and at harvest including the cumulative weed weights (Table 2). In both years, total weed growth as measured by cumulative weed fresh weight was significantly reduced in the weed-free control, when weeds were removed once per week till harvest at 12 WAP compared with the weedy control and the atrazine treatments. The total weight of weeds removed from the weedy control was either similar to or lower than that removed from the atrazine-treated plots in the different years or seasons. Specifically, application of atrazine at 1.0 kg a.i. /ha resulted in significantly higher total weed weight removed compared with the other treatments including the weedy control in the two years, while weed weight removed were similar for atrazine at 2.0 to 4.0 kg a.i. / ha and the weedy control in 2002. A much lighter weed growth was recorded in all the treatments during the late season of 2002 than in the early season of 2003. The atrazine treatments resulted in significantly higher values in 2003. Regressing weed fresh weight (Y) against increasing dose of atrazine (X) showed negative relationships with prediction equations shown in Table 3. The differential weed growth between

Table 2. Effects of atrazine on the cumulative fresh weight (g / m^2) of weeds removed and grain yield (kg / ha) of maize

Treatment	At 4 weeks after treatment		At maize harvest		Total weight of weeds removed		Grain yield	
	2002	2003	2002	2003	2002	2003	2002	2003
Atrazine at 1 kg a.i./ ha	83.1	106.3	271.2	349.0	354.3	455.3	805.5	512.0
Atrazine at 2 kg a.i./ ha	69.3	97.5	210.6	296.4	279.9	393.9	812.5	665.5
Atrazine at 3 kg a.i./ ha	59.3	81.4	223.4	307.2	282.7	388.6	1286.5	1195.3
Atrazine at 4 kg a.i./ ha	62.6	80.1	216.9	277.2	279.5	357.3	1295.8	1200.0
Weedy control	-	-	281.3	340.2	281.3	340.2	675.3	535.8
Weed-free control	-	-	108.0	130.6	108.0	130.6	1302.3	1319.3
S.E. (15DF)	1.60	5.84	4.11	11.23	2.87	5.04	100.24	101.87

treatments was reflected in the significant response in maize grain yield higher in 2002 than 2003 and was found to be positively correlated with increasing dose of atrazine in both years (Table 4). In contrast, correlations of maize grain yield and weed fresh weight were negative in both years (Table 5).

Table 3. Linear correlation and regression analysis between increasing dose of atrazine (X) and weed fresh weight (Y) (n=4)

Weed fresh weight	2002		2003	
	Correlation coefficient (r)	Regression equation	Correlation coefficient (r)	Regression equation
At 4 WAT	-0.88	Y= 86.45-7.15X	-0.96	Y= 115-9.47X
At harvest	-0.70	Y= 268.05-15.01X	-0.87	Y= 358.6-20.46X
Cumulative	-0.78	Y= 354.5-22.16X	-0.94	Y= 473.6-29.93X

Table 4. Linear correlation and regression analysis between increasing dose of atrazine (X) and maize grain yield (Y) (n=4)

Year	Correlation coefficient (r)	Regression equation
2002	+0.90	Y= 563.85+194.49X
2003	+0.94	Y= 244.75+259.38X

Table 5. Linear correlation and regression analysis between weed fresh weight (X) and maize grain yield (Y) (n=4)

Weed fresh weight	2002		2003	
	Correlation coefficient (r)	Regression equation	Correlation coefficient (r)	Regression equation
At 4 WAT	-0.84	Y= 2569.98-22.16X	-0.99	Y= 3436.01-27.84X
At harvest	-0.44	Y= 2082.55-4.48X	-0.70	Y= 3416.2-8.21X
Cumulative	-0.57	Y= 2346.26-4.33X	-0.83	Y= 3758.79-7.19X

Influence of Atrazine on Soil pH and Nutrient Content at Maize Harvest

No consistent trends in soil pH and soil nutrient content were evident amongst the treatments in both years. Significant differences were recorded amongst the treatments in all the soil parameters other than calcium during the late season of 2002 (Table 6). In contrast, significant differences were only noted for soil pH amongst the treatments during the early season of 2003 (Table 7).

In 2002, the weedy control resulted in significantly higher soil pH than the weed-free control and the atrazine treatments, among which there were no significant differences. Organic matter content was significantly higher with the weedy and weed-free controls than with the atrazine treatments. Significant reductions in N content were noted for atrazine at 1.0

and 2.0 kg a.i. / ha compared to the remaining treatments. Atrazine at 3.0 and 4.0 kg a.i. / ha recorded significantly higher values of P than the other treatments. Both K and Mg also varied in a manner more or less consistently related with P for most of the treatments. Regressing soil nutrient parameters (Y) against increasing dose of atrazine (X) indicated positive relationships (Table 8). However, soil pH and increasing dose of atrazine were found to be negatively correlated.

Table 6. Nutrient status and pH of experimental site at harvest in late season maize of 2002

Treatment	pH	Organic matter (%)	N (%)	P (mg / kg)	K (cMol / kg soil)	Ca (cMol / kg soil)	Mg (cMol / kg soil)
Atrazine at 1 kg a.i./ ha	6.2	2.7	0.2	14.5	0.2	1.6	0.6
Atrazine at 2 kg a.i./ ha	6.2	2.9	0.2	13.5	0.2	1.5	0.7
Atrazine at 3 kg a.i./ ha	6.2	2.9	0.3	18.4	0.3	1.7	0.8
Atrazine at 4 kg a.i./ ha	6.1	3.0	0.3	17.9	0.3	1.6	0.8
Weedy control	6.3	3.1	0.3	13.0	0.3	1.4	0.6
Weed-free control	6.2	3.1	0.3	15.3	0.2	1.8	0.8
S.E.(15DF)	0.03	0.10	0.01	0.80	0.01	0.16	0.03

Table 7. Nutrient status and pH of experimental site at harvest in early season maize of 2003

Treatment	pH	Organic matter (%)	N (%)	P (mg / kg)	K (cMol / kg soil)	Ca (cMol / kg soil)	Mg (cMol / kg soil)
Atrazine at 1 kg a.i./ ha	6.2	3.0	0.2	15.0	0.3	1.6	0.6
Atrazine at 2 kg a.i./ ha	6.2	3.0	0.3	14.3	0.3	1.5	0.6
Atrazine at 3 kg a.i./ ha	6.1	2.9	0.3	19.0	0.3	1.5	0.5
Atrazine at 4 kg a.i./ ha	6.1	3.0	0.3	18.2	0.2	1.5	0.5
Weedy control	6.3	3.1	0.3	16.9	0.3	1.6	0.6
Weed-free control	6.3	3.2	0.3	17.2	0.3	1.6	0.6
S.E. (15DF)	0.03	0.11	0.02	1.50	0.01	0.07	0.03

Table 8. Linear correlation and regression analysis between increasing dose of atrazine (X) and soil pH or nutrient content (Y) at maize harvest (n=4)

Soil parameter	2002		2003	
	Correlation coefficient (r)	Regression equation	Correlation coefficient (r)	Regression equation
Soil pH	-0.77	Y= 6.25-0.03X	-0.89	Y= 6.25-0.04X
Organic matter (%)	+0.92	Y= 2.65+0.09X	-0.26	Y= 3-0.01X
N (%)	+0.89	Y= 0.15+0.04X	+0.77	Y= 0.2+0.03X
P (mg / kg)	+0.80	Y= 12.3+1.51X	+0.80	Y= 13.05+1.43X
K (cMol / kg soil)	+0.89	Y= 0.15+0.04X	-0.77	Y= 0.35-0.03X
Ca (cMol / kg soil)	+0.32	Y= 1.55+0.02X	-0.77	Y= 1.6-0.03X
Mg (cMol / kg soil)	+0.94	Y= 0.55+0.07X	-0.89	Y= 0.65-0.04X

As earlier indicated, significant differences were only noted for soil pH amongst the treatments during the early season of 2003. Both the weedy and the weed-free controls recorded significantly higher soil pH values than the remaining treatments, among which no significant differences were indicated. A simple linear correlation and regression analysis between soil pH or nutrient content (Y) and increasing dose of atrazine (X) at harvest showed negative relationships for all the parameters studied, except N and P contents.

Influence of Atrazine Residue from Maize Cropping on Growth and Yield Parameters of Amaranth

Data on growth and yield attributing characters of amaranth as influenced by atrazine residues from maize cropping are summarized in Tables 9 and 10. No statistically detectable differences in the foregoing parameters were indicated amongst the treatments during the late season of 2002. In contrast, significant differences amongst the treatments were noted for all the growth and yield parameters during the early season of 2003, except for leaf area, which remained unaffected. Amaranth from the untreated weedy control plots recorded the lowest values for most of the growth and yield parameters studied. For the most part, the atrazine-treated plots produced amaranth with comparable growth and yield figures to those from the residual weed-free plots. Regressing growth and yield parameters (Y) against increasing residual dose of atrazine (X) showed clearly contrasting relationships in both years. In 2002, all the growth and yield parameters, except leaf number per plant, were negatively correlated with increasing dose of atrazine, while the reverse was the case in 2003 with only leaf number per plant being negatively correlated with increasing dose of atrazine (Table 11).

Table 9. Growth and yield parameters of the common amaranth as influenced by atrazine residues from late season maize in 2002

Residual treatment	Plant height (cm)	Leaf no./plant	Leaf area (cm^2)	Edible yield (g/m^2)	Marketable yield (g/m^2)	Total biomass (g/m^2)
Atrazine at 1 kg a.i./ ha	15.6	9.2	9.0	87.6	87.6	105.5
Atrazine at 2 kg a.i./ ha	13.9	8.9	6.9	73.7	73.7	93.1
Atrazine at 3 kg a.i./ ha	13.5	10.7	6.8	75.8	75.8	93.1
Atrazine at 4 kg a.i./ ha	13.9	9.6	7.8	84.1	84.1	101.0
Weedy control	14.4	9.7	7.7	65.5	65.5	80.0
Weed-free control	15.4	9.2	8.1	81.9	81.9	99.2
S.E. (15DF)	1.33	0.66	1.29	11.15	11.15	9.13

Table 10. Growth and yield parameters of the common amaranth as influenced by atrazine residues from early season maize in 2003

Residual treatment	Plant height (cm)	Leaf no./plant	Leaf area (cm^2)	Edible yield (g/m^2)	Marketable yield (g/m^2)	Total biomass (g/m^2)
Atrazine at 1 kg a.i./ ha	15.2	18.4	10.5	207.4	207.4	249.9
Atrazine at 2 kg a.i./ ha	17.5	18.2	11.0	260.7	260.7	330.7
Atrazine at 3 kg a.i./ ha	20.1	15.0	13.4	278.9	278.9	328.7
Atrazine at 4 kg a.i./ ha	27.4	16.2	13.5	263.6	263.6	323.0
Weedy control	12.0	13.0	14.8	205.9	205.9	266.3
Weed-free control	14.8	16.5	14.9	272.0	205.9	336.9
S.E. (15DF)	0.86	1.03	1.82	6.44	6.44	6.90

Table 11. Linear correlation and regression analysis between increasing dose of atrazine (X) and growth or yield parameters (Y) of amaranth (n=4)

Growth and yield parameters	2002		2003	
	Correlation coefficient (r)	Regression equation	Correlation coefficient (r)	Regression equation
Plant height (cm)	-0.76	Y= 15.6 – 0.55X	+0.96	Y= 10.25 + 3.92X
Leaf number (no./ plant)	+0.49	Y= 8.85 + 0.3X	-0.77	Y= 19.4 – 0.98X
Leaf area (cm^2)	-0.47	Y= 8.55 – 0.37X	+0.94	Y= 9.25 + 1.14X
Edible yield (g/ m^2)	-0.16	Y= 82.40 – 0.84X	+0.77	Y= 205.95 + 18.68X
Marketable yield (g/ m^2)	-0.16	Y= 82.40 – 0.84X	+0.77	Y= 205.95 + 18.68X
Total biomass (g/ m^2)	-0.28	Y= 101.55 – 1.35X	+0.72	Y= 253.75 + 21.73X

DISCUSSION

This study has clearly demonstrated that amaranth is not quite sensitive to atrazine residue from maize cropping and may be sown to succeed maize in rotation without suffering a concomitant reduction in crop growth and yield under rainfed multiple cropping. It also shows that the differential cumulative weed growth and its associated soil pH and nutrient content in the different treatments at maize harvest in both years did not negatively influence the subsequent growth and yield of the rotational amaranth.

The significant differences observed amongst the atrazine treatments in terms of weed weight at 4 WAT and at harvest including the cumulative weed weights indicate that the extent of weed growth in the treatments was dose-dependent. Weed growth decreased more or less significantly in the order of increasing dose of atrazine presumably because higher doses implies a higher quantity of active components available to cause control. This may have accounted for the observed reduction in weed weights and the negative correlations between weed fresh weight and increasing dose of atrazine, particularly at 4 WAT. At harvest, atrazine at all the rates tested was less effective as it failed to provide season-long control of weeds, probably due to its dissipation from the soil environment with time after application, as earlier speculated by Akinyemiju (1992). Akinyemiju *et al.* (1986) indicated that atrazine applied in the humid tropical environment persists for a much shorter period of time after application than in temperate climates. More weed growth occurred in 2003 than in 2002 as a result of relatively higher soil nutrient status and weather conditions that were more conducive to prolific weed growth and greater dissipation of atrazine from the soil during the wet season of the former than the dry season of the latter year. This may have accounted for the relatively higher maize grain yield recorded in 2002 than in 2003.

Significant increases in maize grain yield over the weedy control and the treatments involving atrazine at 1.0 and 2.0 kg a.i. / ha were noted for atrazine at 3.0 and 4.0 kg a.i. / ha and the weed-free control. The differential weed growth between the foregoing treatments was reflected in the significant response in maize grain yield. The highest maize grain yield occurred in the weed-free plots in both years, indicating that it is necessary to protect the crop from weed competition throughout most of its growth to ensure maximum yield. Crop yield might have been increased by hoeing in this study through increased aeration and water movement in the soil (Iremiren, 1988). Of the atrazine treatments tested, atrazine at 3.0 and 4.0 kg a.i. / ha provided the optimum weed control and improved maize grain yield close to that obtained in the weed-free plots, but bearing in mind the economics of labour input and yield, the use of atrazine at the 'standard' recommended rate of 3.0 kg a.i. / ha is still justified and should continue to be encouraged as it is likely to be more economical than the weed-free control because of the prohibitive costs of hiring labour to do manual weeding in the tropics. The poor performance of atrazine at 1.0 and 2.0 kg a.i. /ha in terms of providing adequate weed control and increasing maize grain yield was not quite unexpected as these rates were considered sublethal and their deliberate inclusion in the experimental set-up was done to assess the influence of atrazine residue on rotational amaranth over a sufficient range of atrazine rates. Moreover, it has been established that many herbicides commonly used for weed control show growth-stimulating or hormone-like properties at sublethal concentrations (hormesis) (Akobundu, 1987). This may have accounted for the poor and inadequate weed control by atrazine at the foregoing low dosage rates. Wiedman & Appleby (1972) screened several herbicides in the greenhouse and indicated that many of them from different herbicide groups stimulated plant growth at sublethal concentrations. Plant response to sublethal doses has been extensively reviewed by Ries (1976), who also discussed the possible implication of this phenomenon in agriculture and herbicide rotations (Akobundu, 1987).

In both years, the lack of consistent trends observed in soil pH and soil nutrient content amongst treatments at maize harvest indicates that atrazine exerts very diverse influences on soil nutrient content, thus confirming the conflicting nature of results obtained from previous studies. For instance, several triazine herbicides have been found not to exert any significant influence on soil nutrient content in field trials conducted in established crop rotations (Gruzdyev, 1974). In contrast, significant increases in soil nutrient content over the unsprayed plots have been recorded following repeated applications of triazine herbicides in the field (Kruglov *et al.*, 1975; Sidorov, 1974). The discrepancies in responses of soil nutrient content to atrazine application often reported in the literature appear to be related to, among other factors, differences in soil edaphic and climatic conditions under which these studies were conducted and any lack of uniformity in the physical conduct of the experiments (Lebedeva *et al.*, 1990). In this study, the relatively low soil pH recorded in the weed-free and the atrazine-treated plots at maize harvest presumably arose from water erosion of the exposed soil in the early period of crop growth which led to leaching out of much of the base-forming cations, leaving the exchange complex dominated by aluminum and hydrogen ions (Brady & Weil, 1999). Similarly, the relatively low content of organic matter in the atrazine-treated plots can equally be ascribed to the influence of water erosion on the exposed soil surface that may have led to loss of organic matter. This is consistent with the findings of Gbadamosi *et al.* (2002), who reported similar reductions in organic matter content in atrazine-treated compared to unsprayed plots. The significant reductions in N content recorded in plots sprayed with atrazine at low dosage rates may be the result of poor weed control, which culminated in greater depletion of N by weeds in the affected plots. The much lighter weed growth recorded in the plots sprayed with atrazine at the higher rates presumably contributed to the relatively higher values of P, K and Mg which occurred as a result of reduced uptake of these elements by weeds in the treated plots. Such effects of reduced weed competition on P and the exchangeable cations had earlier been reported for the triazine herbicides (Kruglov *et al.*, 1975; Sidorov, 1974). It has also been argued that the enhanced soil nutrient reserve following triazine application could be as a result of increased microbial activities (Ivanov, 1974; Protasov, 1968), while others deemed it as interplay between reduced nutrient uptake by weeds due to control and increased microbial activities (Sidorov, 1974; Derkach, 1973; Bebekh, 1972). The lack of significant differences in soil nutrient content at maize harvest in 2003 can be attributed to favourable weather conditions that were more conducive to prolific weed growth, resulting in poor growth performance, which tended to minimize the differences among the treatments in terms of soil nutrient uptake.

There were no consistent trends in growth and yield responses of amaranth to atrazine residue in the soil in both years. Significant differences in growth and yield parameters due to soil residual activity of atrazine were not evident amongst the treatments during the late season of 2002, which was not exactly the case in the early season of 2003. Similarly, there were contrasting relationships between increasing dose of atrazine and growth or yield parameters of amaranth in both years. As earlier indicated, in 2002, all the growth and yield parameters, except leaf number per plant, were negatively correlated with increasing dose of atrazine, while the reverse was the case in 2003 with only leaf number per plant being negatively correlated with increasing dose of atrazine. The lack of significant growth and yield differences in amaranth corroborates earlier report by Akinyemiju *et al.* (1986) that atrazine was no longer phytotoxic to tomato seedlings, a related vegetable, 8 weeks after application in the humid tropical environment, which probably explains its inability to provide season-long weed control, presumably due to its dissipation from the soil environment with time after application as speculated by Akinyemiju (1992). The inverse relationships between increasing residual dose of atrazine and growth or yield parameters of amaranth other than leaf number per plant during the late dry season of 2002 suggest that atrazine might possibly impact negatively on the growth performance of this vegetable, probably because of its delayed degradation from the soil environment owing to relatively dry weather conditions. It has been established that atrazine residues in the soil may affect subsequent crops if the interval is too short or if the soil remains dry so that degradation is delayed (Harvey *et al.*, 1982). Conversely, the significant yield increases and the positive relationships obtained with increasing dose of atrazine during the wet season of 2003 suggest faster degradation of atrazine, presumably impacting favourably on the subsequent growth performance of amaranth. This confirms earlier speculation by Walker *et al.* (1982), who indicated that variations in weather pattern between seasons will have a major influence on persistence, because of the effects of temperature and soil moisture content on degradation rates. In general terms, rates of degradation of many herbicides are about 20 times faster in moist soils at 25°C than in dry soil at 5°C (Walker *et al.*, 1982). Thus a warm, moist, well-aerated, fertile soil with optimal pH will be most favourable to microorganisms, which play an important role in the breakdown of atrazine (Klingman & Ashton, 1975).

From the results of this study, it seems evident that atrazine at all the rates tested did not exert any adverse residual effects on the growth and yield of amaranth, suggesting that the vegetable may be adopted to be the rotational crop of choice after a maize crop, in which atrazine has been used for selective weed control without suffering a concomitant reduction in crop growth and yield. However, early cropping season would appear to be better than late season for planting amaranth, because of better prevailing environmental conditions necessary for atrazine degradation.

REFERENCES

Akinyemiju, O.A. (1992). Weed control in maize with acetochlor alone and in combination with atrazine. *Nigerian Journal of Weed Science 5*, 53-62.

Akinyemiju, O.A., Ogunyemi, S.O. & Ojo, I.O. (1986). Persistence of atrazine in a humid tropical soil. *Nigerian Journal of Agronomy 1 (1)*, 22-25.

Akobundu, I.O. (1987).*Weed Science in the Tropics: Principles and Practices*. Wiley. New York

Aladesanwa, R.D. 2005. Screenhouse evaluation of atrazine for soil residual activity on growth, development and nutritional quality of okra (*Abelmoschus esculentus* Moench) in southwestern Nigeria. *Crop Protection 24*, 927-931.

Alvi, S., Perveen, R., Naqvi, I.I. & Shaukat, S. (2003). Effect of atrazine on absorption and translocation of P^{32}, chlorophyll, carbohydrate, protein and potassium contents in bean, *Vigna radiata* (L) Wilczek. *Pakistan Journal of Biological Sciences 6(3)*, 249-251.

AOAC. (1990). *Official Methods of Analysis.* 15th edition. Association of Official Analytical Chemists, Washington, DC, USA.

Bebekh, N.D. (1972). *Primenenie gerbitsidov i desikantov na pocevakh podsolnechnika v usloviakh Omskoi oblasti.* Avtoreferat Kandidatskoi Dissertatsi. Omsk.(In Russian).

Brady, N.C. & Weil, R.R. (1999). *The Nature and Properties of Soils.* 12th Edition, Prentice Hall, Upper Saddle River, New Jersey 07458.

Derkach, A.N. (1973). *Issledovanie dlitel'nosti primeneniia gerbitsidov v zven'iakh ovoshchnogo sevooborota v usloviiakh Altaiskogo Kraia.* Avtoreferat Dissertatsi Kandidata Biologicheskikh Nauk. Barnaul. (In Russian).

Gbadamosi, A.A., Iremiren, G.O. & Aladesanwa, R.D. (2002). Effects of weeding regimes and atrazine application on some soil factors of growth and performance of maize (*Zea mays* L.) in a rainforest area of southwestern Nigeria. *Nigerian Journal of Agriculture, Forestry and Fisheries 3*, 57-60.

Gruzdyev, G.S. (1974). *Primenenie gerbitsidov v sevooborote.* Izdatel'stvo TSKHA. Vypusk 2, 108-121. (In Russian)

Harvey, J.J., Bray, W.E., Jones, A.G. & King, J.M. (1982). Weed control in row crops. In H.A..Roberts (Edited), *Weed Control Handbook: Principles* (7th Edition, pp.292-315). Oxford, Blackwell Scientific Publications.

Iremiren, G.O. (1988). Frequency of weeding for optimum growth and yield of okra (*Abelmoschus esculentus*). *Experimental Agriculture 24*, 247-252.

Ivanov, A.I. (1974). Vlianie simazina na pishchevoi rezhim pochvy. *Agrokhimia 3*, 113-115. (In Russian).

Kauffmann, C., Shoseyov, O., Shpigel, E., Bayer, E.A., Lamed, R., Shosham, Y. & Mandelbaum, R.T. (2000). Novel methodology for enzymatic removal of atrazine from water by CBD-fusion protein immobilized on cellulose. *Environment, Science and Technology 34*, 1292-1296.

Klingman, G.C. & Ashton, F.M. (1975). *Weed Science: Principles and Practices*, Wiley, New York.

Kochhar, S.L. (1986). *Tropical Crops: A Textbook of Economic Botany.* Macmillan Publishers Limited, London and Basingstoke.

Kruglov, Yu. V., Pertseva, A.I. & Galkina, G.A. (1975). Izminenie biologicheskoi aktivnosti pochvy pod vlianiem mnogoletnei sistematicheskoi obrabotki gerbitsidami plantatsii dioskorei kavkazhskoi. *Dokladi VASKHNIL 2*, 20-21. (In Russian)

Lebedeva, G.E., Agapov, V.L., Blagoveshchensky, Yu. N. & Sa.msonova, V.P. (1990). *Gerbitsidi i Pochva: Ekologicheskie Aspekty Primenenia Gerbitsidov.* Izdatel'stvo Moskovskogo Universiteta. (In Russian).

McFarlane, N.R. (1977). *Crop Protection Agents–Their Biological Evaluation.* Academic Press, London.

Protasov, N.I. (1968). Primenenie gerbitsidov (prometrina i alipura) dlia propolki kormovogo liupina. *Sbopnik Nauchnikh Trudov.* Belorusskoi Sel'skokhaistvennoi Akademii. Gorki. *Tom 57*, 84-89. (In Russian).

Ries, S.K. (1976). Subtoxic effects on plants. In L.J. Audus, (Ed.), *Herbicides: Physiology, Biochemistry, Ecology,* volume 2, (2nd Edition, pp. 313-344). London and New York. Academic Press.

Sidorov, V.I. (1974). *Issledovanie dlitel'nogo primenenia gerbitsidov v zvene ovoshchekormovogo sevooborota v usloviakh Moskovskoi oblasti.* Avtoreferat Kandidatskoi Dissertatsi. Moskva. (In Russian)

Walker, A., Briggs, G.G., Greaves, M.P., Hance, R.J. & Thompson, A.R. (1982). Herbicides in soil. In H.A.Roberts (Ed.), *Weed Control Handbook*: *Principles* (7th edition, pp.87-105). Oxford, Blackwell Scientific Publications.

Wiedman, S.J. & Appleby, A.P. (1972). Plant growth stimulation by sublethal concentrations of herbicides. *Weed Research 12*, 65-74.

In: Weeds: Management, Economic Impacts and Biology
Editor: Rudolph V. Kingely
ISBN: 978-1-60741-010-2

Short Communications

LESSONS FROM WEEDS

H. Scholz

Freie Universität Berlin, Zentraleinrichtung Botanischer Garten und Botanisches Musuem, Berlin, Germany

ABSTRACT

The anthropocentric term "weed" has been replaced by "agrestal" and "ruderal." Many agrestal and ruderal taxa are strictly confined to man-made disturbed habitats (farmland and ruderal sites, respectively) and, therefore, must have evolved in the last 12,000 years and still evolve in such places. Among these are anecophytes ("homeless plants", missing in any natural vegetation), anthropogenous taxa and inclusive cultigens. The plurality of evolutionary events resulting in anecophytes demonstrate selected examples, and implications for increasing biodiversity are discussed. Some agrestal and ruderal taxa formerly claimed to be introductions from foreign countries, exotics, are native to regions in which they have were first observed and described. In several cases, the recognition of cryptic anecophytes and cryptic species is decisive.

INTRODUCTION

(For missing literature references, consult the author's papers of 1996 [20] and 2007 [22] and look at the Internet.)

Vascular plants occurring neither sown nor planted in fields, gardens, meadows and pastures, clearing places, along traffic ways, on rubbish heaps in and around towns and cities, in waste lands in industrial and mining areas, on river banks and dikes, in ditches, channels, and so on, commonly called *weeds* and unless tolerated said to be unwanted and harmful, should be more properly named *agrestals* or *agrestalia* (on arable land) and *ruderals* or *ruderalia* (on waste land and in dirty water), to avoid anthropocentric rhetoric. Such plants settling on man-made habitats comprise not only herbs and grasses but also shrubs and trees, often escaped from cultivation and running wild on disturbed ground, such as *Robinia pseudoacacia* L. and *Prunus serotina* Ehrh. (in many parts of Europe introduced from the New World [9]), *Rosa rugosa* Thunb. (in North America and Europe, naturally occurring on shores of NE Asia [13], or, e.g., *Eucalyptus spp.* and *Tamarix spp.* (in tropical-subtropical countries [4]). As from an economic point of view most agrestals and ruderals are sometimes beneficial in preventing soil erosion on mismanaged agricultural and forestry land and initiating restoration of healthy ecosystems, their repeatedly-recommended eradication increasingly gains critical attention. Today, several agrestals are protected under the legacy of nature conservation and biodiversity acts, at least in Germany [9].

The plants of the agrestal-ruderal complex can be divided principally into two groups: apophytes and anecophytes. Apophytes [32] are plant taxa which, with or without human involvement, have left their growing places in natural pristine vegetation and arrived over short or long distances at man-made habitats, or have withstood destruction of their original habitats and survived on the spot and thrived. Anecophytes ("homeless plants," or obligatory "weeds" [20, 22]), in contrast, are anthropogenous taxa strongly confined to man-made habitats, world wide and at any time totally absent in primary or near-natural vegetation. The term "anecophyte" embraces *per definitionem* necessarily also cultigenic crop taxa [25]. The classification is not unambiguous and beyond all dispute in every case, as is the indigenous status of many agrestals and ruderals in a given region [22], and the descent of some anecophytes is obscure (or even unknown if cosmopolitic: "parapophytes" [15], e.g., *Oxalis corniculata* L.).

1. Anecophytic Ruderals

(For convenience, only some taxa are selected.)

In an influential paper on the cytogenetics of brome grasses, Stebbins 1981 [29] on pages 371–72 stated that the Old World genera *Bromus* L. (*Bromus* subgen. *Bromus*) and *Anisantha* K. Koch (*B.* subgen. *Stenobromus* Griseb.), both consisting "entirely of annuals, nearly all of which are ruderal, and many of them strongly weedy, … must have evolved" in the Orient from primitive species of the perennial genus *Bromopsis* Fourr. (*B.* subgen. *Festucoides* Rouy, "*Festucaria*") "to a large extent in adaptation to conditions produced by human agriculture, and particularly the grazing livestock", at the beginning of the Anthropocene [5], the "Neolithic revolution" (V. G. Childe, 1892–1957), about 12,000 years ago. In other words, these annuals are anthropogenous taxa. In later periods spreading to the nearby European and Mediterranean area (and elsewhere), its members went on to further differentiations, resulting in new subspecies and species, mostly polyploids (auto- and allopolyploids) and partly of presumably very recent origin. Striking examples are *Anisantha rubens* (L.) Nevski subsp. *kunkelii* (H. Scholz) H. Scholz (W Mediterraean centred), *Bromus elidis* H. Scholz (Greek endemic), *B. hordeaceus* L. (several regional subspecies), *B. lepidus* Holmberg (W & Central European centred), and *B. nervosus* Acedo & Llamas (endemic of the Iberian Peninsula). Despite molecular research tools widely applied to bromes, the progenitors of the most common polyploid taxa are unknown [23, 24].

The introduction of the genus *Oenothera* L. from the New World started in Europe shortly after 1500 AD. For the imported American species, the old Europe was a virgin "New World" where, on bare and disturbed ground, they had opportunities to multiply, and rapidly diversified. According to Rostański 2006 [16], 62 of the 83 species established at present in Europe are endemics.

Also in the genera *Amaranthus* L., *Aster* L., *Helianthus* L., *Solidago* L., *Xanthium* L., and others, taxa formation in Europe after introduction from America is evident (several regional Floras).

Some ruderals introduced from Eurasia to the New World [7], for instance the annuals *Chenopodium album* L., *Digitaria sanguinalis* (L.) Scop., *Lamium amplexicaule* L. and *Polygonum arenastrum* Boreau (member of the *P. aviculare* L. complex), as well as the perennials *Cirsium arvense* (L.) Scop. (partly agrestal), *Rumex crispus* L. and *Plantago major* L., have no true wild stands in their native Old World; therefore, in the New World, they are non-indigenous anecophytes. In the Mediterranean area, a few taxa (subspecies, varieties and ecotypes) of *Agrostis castellana* Boiss. & Reut. represent indigenous anecophytes (for the situation in Australia compare McLaren 2008 [12]).

2. Anecophytic Agrestals

Since the invention of agriculture in tropical and non-tropical countries, agrestals arose and arise (a) from wild colonisers in adaptation to the continually disturbed farmland habitats, and (b) in various intimate association or interaction with the domesticated cultigenic taxa. [20].

Many agrestals on arable land are anecophytes, predominantly those with short life-cyle, annual or corm- and bulb-forming plants: *Adonis flammula* Jacq., several *Aegilop ssp.* in the Orient, some *Brachiaria spp.* in Africa, infraspecific taxa of the cosmopolitic *Echinochloa crusgalli* (L.) P. Beauv., the northerly distributed Old World taxa of *Buglossoides arvensis* (L.) I. M. Johnst., *Fallopia convolvulus* (L.) Á. Löve and, respectively, *Gagea villosa* (M. Bieb.) Sweet and *Arrhenatherum elatius* subsp. *bulbosum* (Willd.) Schübl. & G. Martens. A few species of the genera *Orobanche* L. and *Striga* Lour. are obligatory parasites on cultigenic host plants. (Examples extracted from the literature and judged by the author.)

In the anecophytic genus *Bromus* (see above), some species and subspecies accompany cultigenic species, cultigens of other genera and mimic them in grain size and specific gravity, loss of dispersal mechanism and quick germination ability: first of all, both the W. European endemics *Bromus bromoideus* (Lej.) Crépin and *B. grossus* Desf. ex DC.; *B. arvensis* subsp. *segetalis* H. Scholz of Central Europe, and the European and W. Siberian *B. secalinus* L. They often infest fields cultivated with the cultigenic cereals *Hordeum vulgare* L., *Secale cereale* L., *Triticum aestivum* L., and *T. spelta* L. (also Eurasian agrestal in wheat fields), but today are in regress through severe control measures. The mimic agrestals can in the large run only reproduce with the help of man and together with the cultigens by joint grain harvesting and sowing (speirochory); otherwise, they vanish. Not so impressive is the connection of *Bromus pseudosecalinus* P. M. Sm. (in Europe) and *B. interruptus* (Hack.) Druce (in England) with breeding strains of some grasses and legumes used for fodder [28].

"Weed-crop mimicry" (Zeven & de Wet 1982 [35]) is also known from other genera. There are anecophytic agrestals in *Agrostemma* L., *Camelina* Crantz, *Galium* L., *Lepidium* L., *Lolium* L., *Persicaria* (L.) Mill., *Silene* L., *Sinapis* L., and *Spergula* L., in distribution strongly confined to fields of cultivated flax (*Linum usitassimum* L.) where some of them mimic the cultigens also in vegetative traits ("vegetative mimicry" and addition to "seed mimicry"); *Lolium temulentum* L., since ancient times a troublesome "weed" in *Hordeum* and

Triticum cultures, and *Echinochlla spp.* in paddy fields (*Oryza sativa* L.) [20]. Unintentionally cultivated, these agrestals are potentially cultigenic. If anecophytic agrestals gave rise to cultigens—well-known examples are "weedy" *Avena* and *Secale spp.* in wheat fields evolving to *A. sativa* L. and *S. cereale* L.—the domestic taxa are called "secondary cultigens." "Primary cultigens" stem directly from non-agrestals or apophytes; anecophytic agrestals do not belong to the ancestry of these cultigens.

Other sources of anecophytic agrestals are cultigenic crop taxa when spontaneously revert back to wild habit and traits. Such "secondary agrestals" arisen by de-domestication processes [6] (from domestic to agrestal taxa, additionally sometimes widespread ruderals), in former times as "primary agrestals" generally but erroneously taken for the progenitors of the crops, are reported from *Cannabis sativa* L. (descendant *C. sativa* subsp. *spontanea* (Vavilov) Serebr.), *Daucus sativus* (Hoffm.) Roehl. (→ *D. carota* L.), *Hordeum vulgare* L. (→ *H. vulgare* subsp. *agriocrithon* (Åberg) Á. Löve), *Lactuca sativa* L. (→ *L. serriola* L. subsp.), *Oryza sativa* L. (→ *O. sativa* var. *fatua* Prain), *Panicum miliaceum* L. (→ *P. miliaceum* subsp. *ruderale* (Kitag.) Tzvelev; *P. miliaceum* subsp. *agricolum* H. Scholz & Mikoláš), *Papaver somniferum* L. (→ *P. somniferum* subsp. *setigerum* (DC.) Coss.), *Pennisetum glaucum* (L.) R. Br. (→ *P. stenostachyum* Klotzsch), the woody *Prunus domestica* L. (→ *P. domestica* subsp. *insititia* (L.) Bonnier & Layens), *Raphanus sativus* L. (→ ?*R. raphanistrum* L. subsp. *frugax* H. Scholz, *nomen provisorium*), *Setaria italica* (L.) P. Beauv. (→ *S. pycnocoma* (Steud.) Henrard ex Nakai), and *Sorghum bicolor* (L.) Moench (→ S. *bicolor* subsp.*drummondii* (Steud.) de Wet & Harlan)—in order only to mention these species, which in parts of their distribution range do not have any true wild relatives, different from the de-domesticated taxa. Therefore, hybridisations between cultigenic taxa and wild taxa, often taken for granted, must be excluded in these cases and fail to explain the existence of "secondary agrestals." Even in cases of co-existence of cultigens and agrestals, introgressive hybridisations, "gene flow," e.g., from *Zea mays* L. to *Z. mexicana* (Schrad.) Kunze in Mexico, and from *Triticum turgidum* L. to *T. dicoccoides* (Körn.) Aarons. in the Near East, is questionable and may indicate cultigens' de-domestications. Surely, hybridisation is an important factor in evolution to anecophytes [3] (for instance, *Oryza* L. in South America [27]) and can combine characters of anecophytic agrestals: *Avena sterilis* L. subsp. × *malzevii* H. Scholz (*A. fatua* L. subsp. *fatua* × *A. sterilis* subsp. *ludoviciana* (Durieu) Gillet & Magne)[19]. Not a "secondary agrestal" is the anecophytic agrestal-ruderal and widely cultivated *Medicago* × *varia* Martyn, product of sexual combination of the cultigenic *M. sativa* L. and the wild *M. falcata* L.

3. Cryptic Anecophytes

The term "cryptic anecophytes" [22] covers all anthropogenous taxa with altered genetic structures (genotypes) not affecting the phenotypes. These hidden gene or genome alterations, traced only by cytological or molecular genetic research techniques (as with GMP), make possible some taxa to spread as cryptic anecophytes over vast areas outside their original homelands in Asia, America and Africa, especially those which often have been named by European authors "archeophytes" or "neophytes" such as *Cichoricum intybus* L. (mostly ruderal on waysides and along roads), *Impatiens parviflora* DC. (ruderal in disturbed forests and shady places), *Galinsoga ciliata* (Rafin.) S. F. Blake and *G. parviflora* Cav. (agrestals of arable land and gardens), and the ruderal *Senecio inaequidens* DC. Several herbicid-resistant agrestals, and some apomicts of the Old World ruderalia of *Taraxacum officinale* Weber aggr., expanded to America, Australia and New Zealand, may also belong to cryptic anecophytes. If two or more cytotypes are recorded in one discrete morphotaxon—the ruderal-agrestal *Bromus commutatus* Schrad. includes diploid (2n = 14) and tetraploid (2n = 28) races—then the recognition of more or less reproductively-isolated "cryptic species" (term coined by zoologists [34], equivalent to "sibling species") is demanded (provided taxa determinations are unequivocal and chromosome numbers are correctly counted). Future investigations can reveal discriminating micromorphological markers perhaps in some cases, and thus remove the representatives from the category of cryptic anecophytes.

Discussion & Conclusion

To adequately grasp the anecophyte concept, one has to realize (1) ongoing evolution up to the present and (2) jumping and rapid taxa formation ("catastrophic selection" [2, 11, 30]) in addition to slow gradualistic processes.

In former times, it was the general belief that all of the plants (and other organisms) occurring in a particular region in man-made habitats are euryoecious apophytes with a wide ecological amplitude tolerating very different and variable conditions. This may be true for casuals and recently-introduced foreign taxa, but not for plants becoming permanently established. Sooner or later, due to time-lag [9], they attain anecophytic status due to a selection-driven adaptation of the plant's genotypes to the novel environment. "The evolution of local adaptation requires genetic variation" [17], and genetic variation within taxa provides a base for larger phenotypic variation and may affect the competitive ability of the ruderals and agrestals. Alternatively to the rigorous adaptation paradigma, currently en vogue in accordance with the neo-darwian Synthetic Theory, the anecophytic ruderals and agrestals in disturbed habitats, *ab initio* largely freed from strong competition and selection forces, may have plaided freely and easily undergone gene or genome combinations ("*patio ludens*"), thus that the individuals with eventually non-adaptive phenotypes (characters) indeed persist in new biotopes ("autonomous evolution" [31]). "A priori not all qualities of a plant must necessarily have survival value; above the survival-value level there is a free scope for qualities (characters) which are irrelevant for its survival" [van Steenis 1969, p. 101, footnote].

Native or introduced?[28]: If in a cultural landscape a new taxon is detected and described, the question often asked is whether or not it is an alien. About 100 years ago, beginning with the studies of the Russian botanist N. W. Zinger on evolution of flax "weeds" (references [20]), continued by N. N. Vavilov and his school on other agrestals, the hypothesis is falsified: new taxa discovered on disturbed ground are introduced from foreign countries (a popular view held even today). The recently described ruderals, partly in press, *Chenopodium lobodontum* H. Scholz (Germany), *Bromus hordeaceus* subsp. *longipedicellatus* L. M. Spalton (W. Europe), *B. incisus* R. Otto & H. Scholz (Germany, Austria), and *B. supernovus* H. Scholz (Australia), are natives, indigenous taxa evolved from related taxa in the concerning regions; introductions did not happen, and proofs that they could be newcomers overlooked or misinterpreted abroad do not exist. Of course, introductions and distribution of taxa, anecophytic or not, occur frequently (topics of invasion biology [10,14, 17]), but such events, at least concerning remote times guessed for "archeophytes,"archeophytes", should not be misused in historical phytogeography and are claimed to be speculative.

Some standing misconceptions around cultigenic taxa may have to be refuted. For example, the woody *Phoenix theophrasti* Greuter, with small fruits, is not an old relict taxon on Crete (Greece), in Turkey and elsewhere, but (very probably) a feral, anecophytic secondary agrestal-ruderal of the cultigenic *Ph. dactylifera* L. [1]. Each cultigen can give birth to a new taxon no longer dependent on human care. Moreover, the so-called introgressive hybrids between cultigenic *Raphanus sativus* L. and agrestal *R. raphanistrum* L. recorded from California (USA) seem to be in reality the introduced Japanese *R. raphanistroides* (Makino) Sinsk. [18], and the ancestor of both the anecophytic agrestal *Avena hybrida* Peterm. and *A. fatua* L. is the cultigenic *A. sativa* L, not the reverse, as is often maintained (see [21]).

Generally disregarded are some authors' opinions on the irrelevance of anthropogenous taxa—especially cultigens—for evolution, by calling this a "non-evolutionary noise" [8].

Despite the inherent problems and irritating instability in taxa ranking and status, and different species concepts used by the authors, the anecophytes are of highest interest for all of the researchers of "anthropogenous florogenesis" [33] because evolution becomes more directly observable than before (evolution in action) and does not rest largely on circumstantial evidence trimmed up in the "tree of life" [26]. Anecophytes are global phenomena, much more so as the human population grows, to c. 6.5 billion people at present (half of them live in cities, nearly the other half on rural lands). Therefore, the human impact on vegetation increases and man's activities diversify the living world in providing the evolution with new impulses and directions, intentionally or unintentionally. How many anecophytes are present in a specific country is difficult to answer; for Germany, the estimates range from 4% to 25% of the total of all species [22]. (Anecozoa are known from flies [*Musca domestica*, diverse *Drosophila spp.*], lice, mice, etc.) Desiderata are intensive taxonomic studies on agrestals and ruderals, mainly in tropical and developing countries.

REFERENCES

[1] Barrow, S. C. (1998). A monograph of *Phoenix* L. (*Palmae*: *Coryphoideae*). *Kew Bull. 53*, 513- 575

[2] Bateman, R. M. & Dimichele, W. A. (1994). Saltational evolution of form in vascular Plants: a neoGoldschmidtian synthesis. In: Ingram, D. S. & Hudson, A. (eds.). Shape and Form in Plants and Fungi. *Linn. Soc. Symposium Ser. 16*, 61-100.

[3] Chapman, M. A. & Burke, J. M. (2007). Genetic divergence and hybrid speciation. *Evolution 61*, 1773-1780.

[4] Coates, P. (2006). *American perceptions of immigrant and invasive species*. Berkely, Los Angeles, London: University of California Press.

[5] Crutzen, P. J. & Stoermer, E. F. (2000). The "Anthropocene". *Global Change Newsletter 41*, 12-13.

[6] Gressel, J. (2005). *Crop fertility and Volunteerism*. Boca Raton, London, NewYork: CRC Press.

[7] 7 Harlan, J. R. (1965). The possible role of weed races in the evolution of cultivated plants. *Euphytica 14*, 173-176.

[8] Hetterscheid, W. L. A. & Brandenburg, W. A. (1995). Culton versus taxon: conceptual issues in cultivated plant systematics. *Taxon 44*, 161-175.

[9] Kowarik, I. (2003). *Biologische Invasionen: Neophyten und Neozoen in Mitteleuropa*. Stuttgart: Eugen Ulmer.

[10] Lambdon, P.W., Pyšek, P., Basnou, C., Hejda; M.., Areianoutsou, M., Essl, F., Jarošik, V., Pergl, J., Winter, Anastasiu, P., Andriopoulos, P., Bazos, I., Brundu, G., Celesti-Grapow, L., Chassot, P., Delipetrou, P., Josefsson, M., Kark, S., Klotz, S., Kokkoris, Y., Kühn. I., Marchante, H., Perglova, I., Pino, J., Vila, M.., Zikos, A., Roy, D., Hulme, P. E. (2008). Alien flora of Europe: species diversity, temporal trends, geographical patterns and research needs. *Preslia 80*, 101-149.

[11] Lewis, H. (1962). Catastrophic selection as a factor in speciation. *Evolution 16*, 257-271.

[12] McLaren, D. A. (2008): Historical aspects of commercial weed invasions – perennial grass case studies. *Plant Protection Quarterly 23*, 65-69.

[13] Mercure, M. & Bruneau, A. (2008). Hybridization between the escaped *Rosa rugosa* (*Rosaceae*) and native *R. blanda* in eastern North America. *Amer. J. Bot. 95*, 597-607.

[14] Prentis, P. J., Wilson, J. R. U., Dormontt, E. E., Richardson, D. M. & Lowe, A. J. (2008). Adaptive evolution in invasive species. *Trends in Plant Science 13*, 288-294.

[15] Ricardo Nápoles, N. E., Pouyu Rojas, E. & Herrera Oliver, P. P. (1996). The synanthropic flora of Cuba. *Fontqueria 42*, 367-429.

[16] Rostański, K. (2006). The occurring of *Oenothera* species in Scandinavia. *Biodiv. Res. Conserv.1-2*, 61-68.

[17] Sakai, A. K., Allendorf, F. W., Holt, J. S., Lodge, D. M., Molofsky, J., With, K. A., Baughman, S., Cabin, R. J., Cohen, J. E., Ellstrand, N. C. McCauley, D. E., O'Neil, P., Parker, I. M., Thompson, J. N. & Weller, S. G. (2001). The population biology of invasive species. *Annu. Rev. Ecol, Syst. 32*, 305-332.

[18] Scholz, H. (1989). Anfrage über Hederich (*Raphanus raphanistrum*). *Flor. Rundbr. 22*, 145- 146.

[19] Scholz, H. (1996). Nachricht über einen hybriden Unkraut-Hafer aus den Südalpen Ober- Italiens. *Ber. Bayer. Bot. Ges. 66/67*, 229-231.

[20] Scholz, H. (1996). Ursprung und Evolution obligatorischer Unkräuter. Origin and evolution of obligatory weeds. *Schriften zu Genetischen Ressourcen. 4* (Festschrift Peter Hanelt), 109-129.

[21] Scholz, H. (2002). Wenig bekannte heimische und fremdländische Gräser Deutschlands (Little known native and alien grasses from Germany). *Flor. Rundbr. 36*, 33-44.

[22] Scholz, H. (2007). Questions about indigenous plants and anecophytes. *Taxon 56*, 1255- 1260.

[23] Scholz, H. (2008). Some comments on the genus *Bromus* and two new species from Europe. *Willdenowia 38* (in press)., 411-422.

[24] Scholz, H. (2008). Die Gattung *Bromus* (*Poaceae*) in Mitteleuropa. Synopse – Synopse und tabellarischer Bestimmungsschlüssel. *Kochia 3* (in press)., 1-18.

[25] Schroeder, F.-G. (2000). Die Anökophyten und das System der floristischen Statuskategorien. *Bot. Jahrb. Syst. 122*, 431-437.

[26] Seberg, O. & Petersen, G. 2007. Assembling the tree of life: Magnitude, shortcuts and pitfalls. In: Hodkinson, T. R. & Parnell, J. A. N. *Reconstructing the tree of life*, 33-46. Boca Raton, London, NewYork: CRC Press.

[27] Second, G. & Rouhan, G. (2008). Human-mediated emergence as a weed and invasive radiation in the wild of the CD genome allotetraploid rice species (*Oryza*, *Poaceae*) in the Neotropics. http://www.plosone.org/doi/pone.0002613

[28] Smith, P. M. (1986). Native or introduced? Problems in the taxonomy and plant geography of some widely introduced annual brome-grasses. *Proceed. Roy. Soc. Edinburgh 89B*, 273- 281.

[29] Stebbins, G. L. (1981). Chromosomes and evolution in the genus *Bromus* (*Gramineae*). *Bot. Jahrb. Syst. 102*, 359-379.

[30] Steenis, C. G. G. J. van (1969). Plant speciation in Malesia, with special reference to the theory of non-adaptive saltatory evolution. *Biol. J. Linn. Soc. 1*, 97-133.

[31] Steenis, C. G. G. J. van (1976). Autonomous evolution in Plants. Differences in plant and animal evolution. *Gardens' Bull., Singapore 29*, 103-126.

[32] Sukopp, H. (2006). Apophytes in the flora of Central Europe. *Polish Bot. Stud. 22*, 473-485.

[33] Tokhtari, V. K. & Wittig, R. (2001). Evolution and development of plant populations in technogenous ecotopes. *Soil Sci. 1*, 97-105.

[34] Winge, H. (1965). Interspecific hybridisation between the six cryptic species of *Drosophila willistoni* group. *Heredity 20*, 9-19.

[35] Zeven, A. C. & De Wet, J. M. J. (1982). *Dictionary of cultivated plants and their regions of diversity*. Wageningen: Centre Agric. Publishing and Documentation.

In: Weeds: Management, Economic Impacts and Biology
Editor: Rudolph V. Kingely

ISBN: 978-1-60741-010-2

Short Communications

EVALUATION OF GLYPHOSATE-BASED PARAQUAT MIXTURES FOR WEED CONTROL IN OIL PALM (*ELAEIS GUINEENSIS*) IN SOUTHWESTERN NIGERIA

R.D. Aladesanwa[1] and E.A. Adegoroye
Department of Crop, Soil and Pest Management,
The Federal University of Technology,
P.M.B. 704, Akure, Ondo State, Nigeria

ABSTRACT

A randomized complete block field trial conducted over 2 years investigated the herbicidal efficacy of glyphosate isopropylamine salt applied singly at 1.44 kg a.e./ ha or in combination with paraquat at 0.2, 0.3, 0.4, 0.5, and 0.6 kg a.i./ha in an oil palm plantation of the Federal University of Technology, Akure, located in the rainforest vegetation zone of Nigeria. Assessment of herbicidal efficacy based on the Henderson-Tilton formula indicated that glyphosate applied alone or in combination with varying concentrations of paraquat proved highly effective in controlling most of the grasses and broadleaved weeds prevalent in the experimental plots. Herbicidal efficacy on total weed density, weed fresh weight, as well as weed dry weight was found to decrease nonsignificantly ($P<0.05$) in the order of increasing concentration levels of paraquat. It was concluded that the use of glyphosate-based paraquat mixtures at the foregoing rates could only be encouraged to improve control of susceptible glyphosate-resistant weed species in an oil palm plantation, where such weed species constitute the main weed population.

***Keywords*:** Glyphosate; Paraquat; Herbicide mixtures; Oil palm

[1] E-mail Address: alarex2003@yahoo.com

INTRODUCTION

Glyphosate and paraquat are by far the most widely used herbicides for weed control in plantation crops in the humid tropics. Glyphosate is a broad-spectrum, nonselective and systemic herbicide that is formulated as an isopropylamine or trimethylsulfonium salt and used for postemergence control of a cross section of annual and perennial weeds including *Imperata cylindrica* (Spear grass), *Panicum maximum* (Guinea grass), and *Cynodon dactylon* (Bermuda grass). Despite its nonselectivity and systemic activity, some differential tolerance has been reported in *Talinum triangulare* (Water leaf), *Cynodon nlemfuensis* (Giant star grass) and *Abutilon theophrasti* (Common ragweed) (Akobundu, 1987; Jordan *et al.*, 1997). Similarly, paraquat is a nonselective, postemergence herbicide but with fast contact action. Its limitation consists in the fact that it is not effective against perennial weeds (Akobundu, 1987).

To obtain good control of heavy infestations of mixed weed populations in the humid tropics, it is often necessary to combine herbicides, since herbicide mixtures are believed to have many distinct advantages over single products, particularly when the mixture components are compatible. Some of these advantages include broad spectrum of action, reduction in the labour required to apply herbicides, reduction in the dosage rates of each component of the mixture applied per unit area of the target, and greater effectiveness resulting from synergistic action of herbicide mixtures.

Therefore, the present study aimed to evaluate the herbicidal efficacy of glyphosate-based paraquat mixtures in an oil palm plantation.

MATERIALS AND METHODS

This study was carried out at the oil palm plantation of the Teaching and Research Farm of the Federal University of Technology, Akure, located in the rain forest vegetation zone of Nigeria during the early (April-July) cropping seasons of 2005 and 2006. The average annual rainfall is about 1300mm with a mean temperature of 27°C and the climate is of the subhumid type. The oil palm plantation was established in 1999 with *tenera* palm seedlings obtained from the Nigerian Institute for Oil Palm Research (NIFOR), Benin-City, Nigeria and planted at the spacing of 8.7 m triangular. Table 1 shows the soil pH and nutrient status at 5 cm soil depth at the experimental sites in both years.

The herbicides used were commercial formulations of glyphosate (Glycel®) containing 41% S.L. of glyphosate in the form of isopropylamine salt, and paraquat (Gramoxone®) containing 276 g of paraquat dichloride. The experiments consisted of eight treatments: (1) a weedy control, where no weeding occurred throughout the experimental period; (2) glyphosate applied alone at the rate of 1.44 kg a.e./ ha (3) paraquat applied alone at 0.6 kg a.i./ ha; (4) glyphosate at 1.44 kg a.e./ ha plus paraquat at 0.2 kg a.i./ ha; (5) 0.3 kg a.i./ ha; (6) 0.4 kg a.i./ ha; (7) 0.5 kg a.i./ ha and (8) 0.6 kg a.i./ ha. Each treatment was replicated four times in a randomized complete block design on individual plots measuring 17 m^2 in area with 1 m alleyway between blocks and 0.5 m between plots. The experiment was set up within the avenues of the palms.

Table 1. Nutrient status and soil pH of experimental sites before treatment application

Year	pH	Organic matter (%)	N (%)	P (mg/kg)	K	Mg	Ca
					(mmol/kg soil)		
2005	6.20	2.84	0.33	10.63	9..25	24.40	35.40
2006	6.15	3.02	0.35	12.34	10.00	25.01	45.04

Preliminary weed assessment was conducted before herbicide application to determine the weed spectrum, weed density, weed fresh weight and weed dry weight of total as well as individual weed species using three fixed 50 cm × 50 cm quadrats along a diagonal in each plot. Subsequently, treatments were applied with a knapsack sprayer fitted with polijet nozzles calibrated to deliver 250 l/ ha of the spray solution at a pressure of 2.5 kg / cm^2. The broadleaved weeds in the plots were already established plants, while the grasses were up to 15 cm high, when spraying was done. Weed control assessment was carried out six weeks after treatment (WAT) by taking weed samples at three sites along a second diagonal in a 50 cm × 50 cm fixed quadrat from each plot. Collected weed samples were bulked, separated by species, counted, oven-dried at 80 °C for 48 hours and subsequently weighed.

Percentage herbicidal efficacy on total as well as on individual weed populations including their fresh and dry weights was determined using the Henderson-Tilton formula (Puntener, 1981) based on nonuniform weed infestation in the plots before application.

Collected data were submitted to an analysis of variance (ANOVA) and the Duncan's multiple range test (DMRT) was used to verify the significant differences among treatment means at the 5% probability level. Data on weed density and percentage herbicidal efficacy were normalized prior to ANOVA using the square root and arcsine transformations, respectively.

RESULTS AND DISCUSSION

Means of the analyzed pooled data for the different years are presented in the tables, since total as well as individual herbicidal efficacies on weed growth in both years showed similar trends and indicated no statistically detectable differences. Assessment of total herbicidal efficacy based on the Henderson-Tilton formula showed that there were significant ($P<0.05$) differences amongst treatments in terms of percentage weed control either by density, weed fresh weight, or weed dry weight, respectively (Table 2). Glyphosate applied singly was better than any of the treatments based on weed control by density, although there were no significant differences between it and most of the treatment combinations. Glyphosate plus paraquat at 0.5 kg a.i. / ha or at 0.6 kg a.i. / ha recorded the highest percentage weed control by fresh weight or dry weight, respectively. Glyphosate is a broad-spectrum, nonselective and systemic herbicide that is used for post emergence control of both annuals and perennial weeds (Akobundu, 1987). Specifically, glyphosate is an inhibitor of the enzyme 5-enolpyruvylshikimate-3-phosphate synthase which is involved in aromatic amino acid biosynthesis (Cole, 1982). Paraquat is also a nonselective post emergence herbicide with fast contact action. It interferes with photosynthesis by diverting electrons from ferrodoxin during the non-cyclic electron flow (Akobundu, 1987).

Table 2. Mean total percentage herbicidal efficacy at 4 weeks after treatment application (2005-2006).

Treatment	% control by density	% control by fresh weight	% control by dry weight
Glyphosate only	74.97a	89.78ab	91.36a
Paraquat only	61.68ab	90.14ab	86.38a
Glyphosate + paraquat at 0.2 kg a.i./ ha	55.78b	87.36ab	91.54a
Glyphosate + paraquat at 0.3 kg a.i./ ha	68.22ab	81.28b	90.29a
Glyphosate + paraquat at 0.4 kg a.i./ ha	67.20ab	88.92ab	92.58a
Glyphosate + paraquat at 0.5 kg a.i./ ha	66.55ab	93.4a	93.81a
Glyphosate + paraquat at 0.6 kg a.i./ ha	66.39ab	90.46ab	93.86a
Weedy control	0.00c	0.00c	0.00b

Means in a column followed by the same letter are not significantly different by DMRT ($P<0.05$)

Herbicidal efficacy on individual weed species based on weed density varied amongst treatments (Table 3). Most of the weeds prevalent in the plots were effectively controlled by most treatments. Percentage weed control by density varied from 50 to 100%. *Sporobolus pyramidalis*, *Axonopus compressus*, and *Centrosema pubescens* were controlled 100%. Most of the treatments proved highly effective in reducing weed populations in the plots. However, the combination of glyphosate with paraquat at 0.4 kg a.i. / ha and above caused an antagonistic effect on the control of *Aspilia africana* and *Melanthera scandens*. The antagonistic reduction in the activity of glyphosate when combined with growth regulator herbicides, such as bromoxynil and 2, 4-D has been reported by Akobundu (1987).

Overall, the results of this study suggest that the use of herbicide mixtures involving glyphosate and paraquat at 0-3 kg a.i./ha could only be encouraged to improve control of susceptible glyphosate-resistant weed species in an oil palm plantation, where such weed species constitute the main weed population. Alternatively, it may be desirable to apply these herbicides sequentially to optimize their efficacies rather than mixing them because of their suspected chemical incompatibility, which presumably resulted in the loss or reduction of effectiveness of one or both components.

Table 3. Mean herbicidal efficacy (%) on individual weed species based on percent control by density (2005-2006).

Treatment	*A. aspera*	*A. africana*	*A. compressus*	*C. pubescens*	*C. diffusa*	*I. batatas*	*M. scandens*	*S. pyramidalis*	*T. triangulare*
Glyphosate only	100.00a	100.00a	100.00a	100.00a	100.00a	83.00b	100.00a	100.00a	83.00b
Paraquat only	100.00a	100.00a	100.00a	100.00a	100.00a	50.00c	100.00a	100.00a	100.00a
Glyphosate +paraquat at 0.2 kg a.i./ ha	100.00a	100.00a	100.00a	100.00a	50.00b	100.00a	100.00a	100.00a	100.00a
Glyphosate + paraquat at 0.3 kg a.i./ ha	100.00a	95.00a	100.00a	100.00a	100.00a	100.00a	100.00a	100.00a	100.00a
Glyphosate + paraquat at 0.4 kg a.i./ ha	100.00a	70.13b	100.00a	100.00a	100.00a	100.00a	83.33b	100.00a	100.00a
Glyphosate + paraquat at 0.5 kg a.i./ ha	99.00a	65.14b	100.00a	100.00a	100.00a	100.00a	96.40b	100.00a	100.00a
Glyphosate + paraquat at 0.6 kg a.i./ ha	100.00a	64.71b	100.00a	100.00a	100.00a	100.00a	96.60b	100.00a	100.00a
Weedy control	0.00b	0.00c	0.00b	0.00b	0.00b	0.00d	0.00c	0.00b	0.00c

Means in a column followed by the same letter are not significantly different by DMRT ($P<0.05$)

REFERENCES

Akobundu, I.O. (1987). Weed Science in the tropics: Principles and Practices. New York. Wiley.

Cole, D.J. (1982). The mode of action of glyphosate. British Crop Protection Conference-Weeds, pp.309-315.

Jordan, T.N., York, A.C., Griffin, J.I., Clay, P.A., Vidrine, P.R. & Reynolds, B. (1997). Influence of application variables on efficacy of glyphosate. Weed Technology 11, 354-362.

Puntener, W. (1981). Manual for field trials in plant protection. Basle, Switzerland, Ciba-Geigy Limited.

In: Weeds: Management, Economic Impacts and Biology
Editor: Rudolph V. Kingely
ISBN: 978-1-60741-010-2

Short Communications

WEED MANAGEMENT WITH NATURAL PRODUCTS: PROSPECTS OF FUNGAL PHYTOTOXIN RESEARCH

Maria Chrysayi-Tokousbalides*

Pesticide Science Laboratory, Agricultural University of Athens,
75 Iera Odos, 118 55 Athens, Greece.

ABSTRACT

Concern on the safety of agrochemicals has led to more stringent international and national regulations for registration of new protection products. In an effort to maximize the chances of developing effective products with acceptable performance regarding environmental and human safety, the emphasis has been given on discovering and screening new active compounds. The input to those screens is derived from laboratory synthesized compounds but also from natural extracts with some kind of bioactivity. As far as the development of new herbicides is concerned, research on natural phytotoxic compounds has given novel chemical templates and sites of action. Several experimental protocols have been developed by research groups in order to isolate, characterize, evaluate the bioactivity and study the mode of action of natural phytotoxins. Natural phytotoxic compounds are mainly secondary metabolites of fungi or plants and are characterized by structural complexity and diversity. A large number of fungal phytotoxins have been identified and have exhibited interesting selective phytotoxicity. Unfortunately, such compounds have not fully been exploited for the development of new herbicides due to the difficulties associated with the multidisciplinary character of natural products research. Several bioactive natural compounds are structurally too complex for synthesis and low cost production. Such complex molecules do not facilitate delivery to the test-organism which is important for proper *in vitro, in vivo* and *in planta* screening. In order to overcome such obstacles research must be focused on the development of case-specific protocols and bioassays for evaluation of the herbicidal potential and elucidation of the mode of action. Other important aspects such as chemical properties, persistence and selectivity at the organism level must be examined before considering a phytotoxin as a lead compound for herbicide development. Fungal phytotoxins must be subjected to the same tests and regulation restrictions applied to

* Correspondent Author: mchrys@aua.gr

synthetic phytotoxic molecules. Since the most fungal phytotoxins are produced in small quantities by phytopathogenic fungi and cause metabolic changes to affected plants, recent approaches such as genomics and metabonomics will be valuable tools in phytotoxin research.

INTRODUCTION

Public concern on the adverse effects of chemicals on humans and the environment has caused sufficient pressure for enforcement of more stringent legislation. Since agricultural practices involved in crop management and protection rely on agrochemical use, relevant acts, directives and guidelines have been introduced in order to register new or re-evaluate older plant protection products. Bioactivity evaluation, product specifications, delivery technologies, toxicological issues and environmental fate are now more studied than in the past. Major attention has been given to consumer risk assessment and environmental impact of products under development. Compounds of high selectivity towards target organisms but also of low persistence are considered to be of lower risk to humans and ecosystems. As a result the rationale behind the discovery and development of pesticides has been modified and it is more difficult for agrochemical companies to develop and take into market new products. In an effort to maximize the chances of discovering new active ingredients with acceptable performance regarding environmental and human safety emphasis has been given on increasing the number of substances tested. The input to compound screening is currently derived from several sources including laboratory synthesized compounds and natural extracts which generally contain bioactive compounds with interesting structural differences than the synthetic ones [35]. As far as the development of new herbicides is concerned, major consideration has been given to the investigation of phytotoxic substances of natural origin, such as phytotoxins produced by fungal pathogens [24, 25]. The biosynthesis of phytotoxins by the parasite is often essential for the infection onset and further host colonization [72]. Such fungal metabolites are characterized by various degrees of selectivity and chemical diversity [36, 37, 60]. Therefore, a large number of fungi which are plant pathogens have been isolated from infected plant tissues, cultured on growth media and the herbicidal potential of their metabolites has been examined. In addition, a number of known fungal phytotoxins have been synthesized in the laboratory and their bioactivity has also been studied. The challenge lies not only on the optimization of their properties or on the discovery of novel templates, but also on the evaluation of the degree and basis of their selectivity at the organism level.

Since ecotoxicological aspects are also of concern for the development of new herbicides, studies on the bioactivity of the phytotoxins under consideration must be directed not only to model weed species but towards non-target organisms. Major area of concern is, therefore, the selection of test species for studying the bioactivity and the selective toxicity of the compounds. It is also important to evaluate the responses of the selected organisms by taking proper observations and measurements. The obtained qualitative and quantitative data might also provide information important for the orientation of mode of action studies. Taking into account that so far there has been little overlap between target sites of synthetic herbicides and fungal phytotoxins [3, 4, 7, 15, 19, 20], phytotoxin research would be really interesting and quite often rewarding.

This chapter is meant not to be exhaustive but rather to suggest and define areas of research which may permit further utilization of fungal pathogens and their metabolites. The evaluation of bioactivity of fungal phytotoxins and the selection of promising lead compounds have been achieved independently by several research groups through different and not directly comparable methodology. Such lack of harmonization permits judgment on a case to case basis. For the purpose of this chapter the fungal phytotoxins pyrenophorol (5,13-dihydroxy-8,16-dimethyl-1,9-dioxa-cyclohexadeca-3,11-diene-2,10-dione) and pyrenophorin (8,16-dimethyl-1,9-dioxa-cyclohexadeca-3,11-diene-2,5,10,13-tetraone) have been selected as examples (Fig. 1). They belong to macrodiolides and they are both produced in cultures of a *Drechslera avenae* pathotype which is a host-specific pathogen of *Avena sterilis*. Based on available data on the above phytotoxins, the following key steps in fungal phytotoxin research will be presented and discussed: Isolation of phytotoxins, phytotoxicity, mode of action studies and activity on non-target organisms. The examples given are neither all that they exist, nor do they necessarily represent those that will eventually find their way to the market.

PHYTOTOXIN ISOLATION

Research on natural phytotoxins has a multidisciplinary character but the ultimate aim of all studies is to evaluate the bioactivity of a compound having a particular chemical structure. Therefore, isolation and characterization of metabolites are requisites for successful structure-activity studies. The variation of fungal metabolites is enormous and differs with the fungal strain, stages of fungal development and conditions of growth. The selection of the fungal strain is very important because not all isolates are efficient phytotoxin producers *in vitro*. The approach we have used in our laboratory in order to study phytotoxins of fungal origin is dictated by the apparent role of the compound in nature. Therefore, we have focused our efforts on metabolites produced by fungi which are non-obligate, host-specific plant pathogens of high virulence. The virulence is tested with pathogenicity tests and then the most virulent and host-selective strains are cultured on a series of growth media. Culture fractionates are bioassayed for phytotoxicity symptoms on detached leaves or seedling cuttings of host-plants. Only fractions which cause physiological and morphological responses similar to disease symptoms are further purified and the phytotoxic agent is isolated and structurally characterized with appropriate methodology and instrumentation.

The composition of the growth medium is a determining factor and must be taken into consideration before discarding isolates as non or poor toxin-producers. In the case of the pathotype of *D. avenae* which was used in our experiments, the filtrates from cultures on oatmeal agar gave extracts with the highest phytotoxicity to *A. sterilis*. Such observation lead to further purification of the extracts and to the isolation of (5*S*,8*R*,13*S*,16*R*)-(-)-pyrenophorol and (8*R*,16*R*)-(-)-pyrenophorin. Medium similar to the natural substrate [63] or minimal media have better served phytotoxin production depending on the fungus and conditions of growth. Also, the temperature is an important parameter for maximization of phytotoxin production. The optimal temperatures for obtaining relatively high amounts of (5*S*,8*R*,13*S*,16*R*)-(-)-pyrenophorol and (8*R*,16*R*)-(-)-pyrenophorin are within the range of 15-20 °C [42, 43]. Phytotoxins as products of secondary metabolism are produced during the stationary stage of growth but in order to harvest at the maximal yield monitoring of the production of each phytotoxin is necessary. The most phytotoxins reach a maximum yield in 3-4 weeks after medium inoculation but a few phytotoxins, as in the case of pyrenophorin, can be detected 3 days after inoculation reaching a maximum yield 8 days later [43]. Increasing incubation temperature to 20 °C stimulated phytotoxin production by *Alternaria radicina* [63]. Special manipulations were required in order to obtain acceptable yields of the host-selective phytotoxin phomalide from cultures of virulent strains of *Phoma lingam* [56]. The phytotoxin produced by the fungus was phomalide and the yield of the metabolite was easily affected by the age and density of spores in the inoculum. Also the presence of sirodesmin PL, which is produced by the mycelium and not by the spores, interferes with phomalide production [56].

A large number of phytotoxins had been isolated and chemically characterized long before their activity has been examined, as in the case of the metabolites pyrenophorol and pyrenophorin. Pyrenophorol was isolated for the first time in 1969 from cultures of *Byssochlamys nivea* [45] and later it was detected in cultures of *Stemphylium radicinum* [33]. Pyrenophorin was first isolated in 1961 from *Drechslera avenae*, a pathogen of cultivated oats [40] and later from *Stemphylium radicinum* [33] and *Pyrenophora avenae* [47]. Both phytotoxins have been produced by total synthesis [23, 48, 51]. Pyrenophorol has four and pyrenophorin two stereoisomers. Although the chemistry of the metabolites had been known for decades limited and fragmentary information on their bioactivity had been reported. During the process of our research on phytotoxic fungal metabolites of the particular *D. avenae* pathotype we were able to isolate only the stereoisomers (5*S*,8*R*,13*S*,16*R*)-(-)-pyrenophorol and (8*R*,16*R*)-(-)-pyrenophorin (Fig. 1) which were not detectable in cultures of avirulent *D. avenae* isolates.

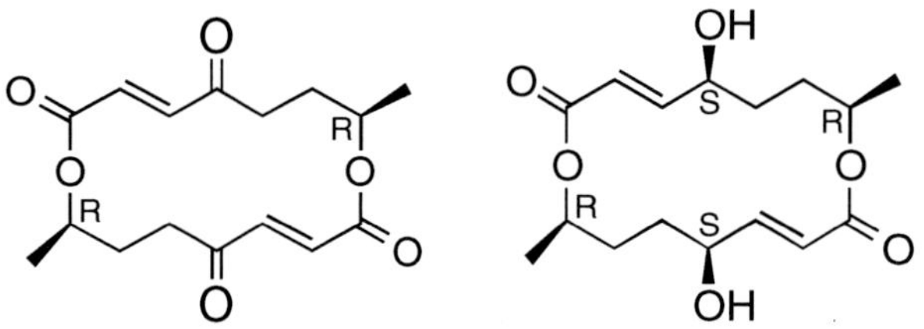

B

Fig. 1. Structures of the phytotoxins (8*R*,16*R*)-(–)-pyrenophorin (left) and (5*S*,8*R*,13*S*,16*R*)-(–)-pyrenophorol (right).

The development of sophisticated methodology and instrumentation has facilitated purification and chemical characterization of fungal metabolites. The need for synthesis still arises though, because in most cases there is insufficient quantity of the metabolite to make extensive studies on bioactivity, selectivity range, stability, uptake and translocation within biological systems. In spite of the fact that the most bioactive natural compounds are structurally too complex to be synthesized, total synthesis has been achieved for a number of fungal phytotoxins [8-10, 18, 26, 34, 38, 46, 48, 50, 62, 72]. Synthesis of fungal metabolites in the laboratory is important for covering the need for large quantities not only of the isolated natural metabolite but also of various chemical analogs which are required for structure-activity studies [10, 34, 72]. In addition synthesis of enantiomers is important since differences in the bioactivity of isomers have been observed [9, 38]. The phytotoxin phomalide is active *in planta* at 10^{-5} M but its (*Z*)-isomer was not found phytotoxic possibly due to isomerization in the plant [72].

The genetic control of phytotoxin production by plant pathogenic fungi has been considered as a tool for studying not only pathogenicity mechanisms but also the biosynthesis of the phytotoxins as well [11]. Development of methods for the genetic manipulation of phytotoxin producing fungi would certainly contribute to the maximization of phytotoxin production in culture.

PHYTOTOXICITY AND MODE OF ACTION STUDIES

Since there are not established bioassay protocols, several experimental methods have been developed by research groups in order to study the response of plants when exposed to fungal phytotoxins. Regardless the biochemical, physiological or morphological approach of the work, common denominator to all phytotoxicity studies is the use of plant tissues as test material. The most research groups use as test plants for the initial phytototoxicity screening the hosts of the fungal pathogens from cultures of which the phytotoxins have been isolated. Other weed, crop or model species have been included in selectivity or mode of action studies [2-5, 20, 28, 34, 41-43, 53, 68]. *In vitro, in vivo* and *in planta* assays have been employed for evaluating the herbicidal potential of various test preparations and for investigating the mode of action of phytotoxins. The selected plant material varies, so does the method of application. Test solutions have been applied on leaves on intact seedlings, stem cuttings, detached leaves, leaf segments, pre-germinated seeds or ungerminated seeds [1-6, 10, 27-32, 41-43, 52-54, 57, 68]. The phytotoxin preparations could also be applied internally through the vascular system by immersion of stems or seedling cuttings or by injecting the solution to cortical tissue [30-31, 43]. External exposure of plants or plant parts could be made by drop, spray or brush application. The mode of application determines the efficiency of deposition and uptake of test solution. The phytotoxin cyperin inhibited root growth when the seedlings were grown on agar but the phytotoxin was inactive when applied on soil or foliage at greenhouse conditions [34]. Another important issue is to correlate the nominal with the real concentration taken up by the plant tissue. In the case of immersion bioassays a way to manipulate this problem is to maintain the seedlings, stem cuttings or excised leaves in the preparation until the entire solution is taken up and then maintain the plant parts into distilled water until the recording of symptoms of phytotoxicity [29, 54]. Also the plant part might influence the expression of phytotoxicity. Maculosin was inactive on whole plants. Young leaves or stem tips were found tolerant to synthetic maculosin analogs. [10].

Phytotoxins as complex molecules do not facilitate delivery to the test-organism which is very important for proper screening. The structural differences play an important role not only on phytotoxicity but possibly on uptake and translocation in the plant as well. The differences in plant uptake and translocation observed between pyrenophorol and pyrenophorin led us to apply different protocols for studying phytotoxicity and the mode of action of the phytotoxins. The fact that pyrenophorin was not taken up by *A. sterilis* seedlings was a limiting factor for not applying the metabonomics approach for studying the mode of action of this phytotoxin and the method was applied successfully for the investigation of the mode of action of pyrenophorol [4].

Lack of phytotoxicity might be totally or partially attributed to low uptake and translocation of the phytotoxin and not to the insensitivity of the target site itself. Since the polarity of most phytotoxins does not permit cuticle penetration, in several cases lesions or punctures were made on the leaves in order to facilitate phytotoxin uptake [10, 27, 32, 42, 52, 54, 57]. A way to facilitate plant penetration is to use small amounts of proper organic solvents or adjuvants in combination with water. Also plant cell suspension cultures have been employed for delivering of phytotoxins to plant cells without the obstacle of cuticle [2, 54]. The toxicities of pyrenophorol and pynenophorin to tissue cultures of *A. sterilis* and *A. sativa* were comparable to those exhibited by the phytotoxins on leaves or cuttings of the respective species [2]. Such finding indicates that the basis of selectivity of pyrenophorol and pynenophorin is associated with cellular functions and is not related to differences in cuticle penetration.

After selecting a suitable plant tissue and delivery method, experiments are needed to obtain measurements for making a dose-response curve. Regardless the type of the assay and determined parameters [20], it is very useful for directly made comparisons that concentration of the phytotoxin to be reported in molarity units. Differences in the expression of phytotoxicity of pyrenophorol seem to depend on specific stereochemistry, as in cases of several herbicides. The fungal metabolite pyrenophol [65] with a different stereochemical configuration from (5*S*, 8*R*, 13*S*, 16*R*)-(-)-pyrenophorol was not found phytotoxic to a range of plants tested. On the other hand, a compound isolated from *Alternaria alternata* and named helmidiol, which is identical to (5*S*, 8*R*, 13*S*, 16*R*)-(-)-pyrenophorol, has anthelmintic activity [44].

Although the use of empirical index or arbitrary scale in phytotoxicity bioassays give an estimate of activity, the data might be of a higher value if they are taken with a more quantitative method such as chlorophyll content determination, protein content or conductivity measurements (3, 43]. Such measurements would also be a valuable contribution to the mode of action studies of the compound.

Activity on Non-Target Organisms

Since fungal phytotoxins can be exploited for the development of new herbicides, more information about the effects on non-target organisms needs to be known. The data produced through toxicity bioassays on non-target organisms could be used as a base for further ecotoxicological studies in a technical scale. Also the responses of organisms belonging to different taxa would be additional physiological clues of subcellular targets of the phytotoxins. Since aquatic ecosystems could be affected by plant protection products, model organisms such as the marine bacterium *Vibrio fischeri,* the cyanobacterium *Oscillatoria perornata*, the green alga *Pseudokirchneriella subcapitata*, the aquatic plant *Lemna minor,* the crustacean *Artemia fransiscana* and the mosquitofish *Gambusia affinis* were used in bioassays for testing the bioactivity of pyrenophorol and pyrenophorin.

The Microtox® assay, which employs the marine bacterium *Vibrio fischeri*, was used to examine the effects of both phytotoxins on a model prokaryote. Pyrenophorin had the lowest EC_{50} values (3.57 and 0.97 M 10^{-5}, at 5 and 15 min of exposure) exhibiting higher activity than those of diuron, glyphosate, mesotrione, norflurazon, oxadiazon or paraquat [17]. On the contrary, pyrenophorol (with corresponding EC_{50} values of 801 and 400.64 M 10^{-5}) was much less toxic than pyrenophorin and exhibited toxicity at the same order of magnitude with the above herbicides.

In regards to cyanobacterial sensitivity, pyrenophorin was found to be more toxic towards *O. perornata* than pyrenophorol [17]. The cyanobacterium *O. perornata* was more sensitive than the green alga *P. subcapitata* to pyrenophorin. Also *P. subcapitata* exhibited a similar lack of sensitivity to both phytotoxins [17].

Duckweed species (*Lemna* spp.) are widely used in phytotoxicity bioassays [49] and are suggested as bioindicators [66, 39]. In our tests, neither pyrenophorol or pyrenophorin was found as toxic to *L. minor* as the herbicide paraquat. The toxicity of pyrenophorin was found to be at the same order of magnitude with that of the weak uncoupler 3,5-dichlorophenol and much more toxic than pyrenophorol. Also pyrenophorol was less active on roots than on fronds of *L. minor* [17]. The response of *L. minor* to several phytotoxins produced by *Fusarium moniliforme* has also been examined. The phytotoxin fumonisin B1 was found to be the most active [69].

In the addition to the above aquatic organisms, more complex life forms like the larvae of the crustacean *Artemia fransiscana* (formerly *Artemia salina*) and adults of the mosquitofish *Gambusia affinis* were used in our acute toxicity bioassays. The phytotoxin pyrenophorin exhibited higher acute toxicity on *A. fransiscana* larvae with LC_{50} value of 14.7 M 10^{-5} than pyrenophorol (LC_{50} of 64 M 10^{-5}). It has been reported that other fungal phytotoxins such as eremophilanes, esters of fusaric or 9,10 dehydrofusaric acid, drazepinone and trichothecenes are toxic to *A. salina* at concentrations varying from 10^{-8} to 10^{-3} M [1, 6, 12-14, 28].

In regards to acute fish toxicity, pyrenophorin was more toxic than paraquat to *Gambusia affinis,* but not as toxic as malathion [67]. The data summarized in Table 1 show that the ecotoxicological profile of pyrenophorol is safer than that of pyrenophorin. The phytotoxicity of (8*R*,16*R*)-(-)-pyrenophorin on *A. sterilis* has been associated with plant processes related to oxidative stress and the production of reactive oxygen species [3, 5]. Such mode of action might be responsible for the toxicity of the phytotoxin on a wide range of organisms (Table 1). The observed differences in the expression of toxicity between pyrenophorol and pyrenophorin is indicative of existed differences in the availability and/or mode of action of the phytotoxins.

The generalization that natural products are not persistent and environmentally safe is an assumption which must be proved true only by proper experimentation. Studies on the potential of phytotoxic extracts from cultures of *Phoma sepium* f. sp. *arvensi* to be exploited for the development of new herbicides against *Convolvulus arvensis* showed that phytotoxic culture filtrates retained 50% and 25% of their initial bioactivity after subjected to boiling for 15 and 30 minutes, respectively [16]. The environmental fate of fungal phytotoxins has not yet been extensively studied. It is encouraging that a methyl ester of fusaric acid and its degradation product are not persistent in three soil types with half life values maximum 41-45 days at 10 °C [70].

Table 1. Comparative selectivity of the fungal phytotoxins (5*S*,8*R*,13*S*,16*R*)-(-)-pyrenophorol and (8*R*,16*R*)-(-)-pyrenophorin on plants and aquatic organisms [43, 17, 67].

Test organism	Bioassay	Toxicity	
		Pyrenophorol (M 10^{-5})	Pyrenophorin (M 10^{-5})
Avena sterilis	seedling cutting	64^a	$-^b$
	seed germination	$-^b$	7^a
	root growth	$-^b$	7^a
Avena fatua	seedling cutting	64^a	$-^b$
	seed germination	$-^b$	7^a
	root growth	$-^b$	7^a
Avena sativa	seedling cutting	$-^b$	$-^b$
	seed germination	$-^b$	7^a
	root growth	$-^b$	7^a
Triticum aestivum	seedling cutting	$-^b$	$-^b$
	seed germination	$-^b$	7^a
	root growth	$-^b$	7^a
Hordeun vulgare	seedling cutting	$-^b$	$-^b$
	seed germination	$-^b$	7^a
	root growth	$-^b$	7^a
Vibrio fischeri	bioluminescence	801^c	3.57^c
Oscillatoria perornata	inhibition	$>10^d$	1.3^d
Pseudokirchneriella subcapitata	inhibition	$>10^d$	$>10^d$
Lemna minor	frond discoloration	287^c	6.10^c
	rhizoid number	3059^c	4.53^c
Artemia fransiscana	nauplii mortality	64^e	14.7^e
Gambusia affinis	adult mortality	NT^f	2.6^e

[a]Minimum effective concentration. [b]Non phytotoxic. [c]Median effective concentration (EC_{50}). [d]Median inhibitory concentration (IC_{50}). [e]Median lethal concentration (LC_{50}). [f]Non-tested.

Within this context, the acute toxicity of chemicals which are lead candidates must be examined on organisms belonging to different trophic levels. In the majority of ecosystems the most important trophic level is occupied by procaryotes, thus it is important to include bacteria in the series of assays for the evaluation of phytotoxins. We have found the Microtox® applicable in phytotoxin research as it has been found in toxicity assessments of other aquatic pollutants [59, 64]. Algicide tests using green algae and cyanobacteria have been valuable for testing toxicity effects on phytoplankton organisms [61] and could also be successfully applied in phytotoxin evaluation. The use of bioindicators such as duckweed species (*Lemna* spp.), the crustacean *Artemia fransiscana* and the mosquitofish (*Gambusia spp.*) [58] has been employed in various studies and could be a valuable tool in phytotoxin toxicity screening. It is also worth mentioning that it is important to include reference chemicals in the experiments. The inclusion of phytotoxic compounds with known in practice performance would give a better evaluation of the bioactivity of a phytotoxin.

CONCLUSION

Nature has very successfully preceded man in synthesizing chemicals. The potency and properties of chemicals which are produced in nature's laboratories must be studied in depth before they are handled in bulk. Regardless of source all phytotoxic compounds must be subjected to the same restrictions and regulations before they are introduced by man into the environment. Data on the selectivity of the phytotoxins (5*S*,8*R*,13*S*,16*R*)-(-)-pyrenophorol and (8*R*,16*R*)-(-)-pyrenophorin show that fungal phytotoxins have the potential to affect a range of organisms not necessarily involved in the evolution of the phytotoxin producing fungus. As a result, research on fungal phytotoxins is of multidisciplinary character and has been associated with problems of variable nature. In order to overcome such obstacles research must be focused on the development of case-specific protocols for optimization of phytotoxin isolation, chemical characterization, phytotoxicity evaluation and mode of action studies. As far as the environmental safety is concerned, aspects such as chemical stability, water solubility, persistence and selectivity at the organism level must be studied before considering a phytotoxin as a lead compound for herbicide development.

REFERENCES

Abouzeid, M., Boari, A., Zonno, M. C., Vurro, M. & Evidente. A., (2004). Toxicity profiles of potential biocontrol agents of *Orobanche ramosa*. Weed Sci. 52, 326-332.

Aliferis, K. A., & Chrysayi-Tokousbalides, M., (2001). The use of cell suspension cultures of *Avena sterilis* L. for the study of herbicidal activity of fungal metabolites. *Abstracts, International Symposium on Bioactive fungal metabolites-Impact and Exploitation*, 42

Aliferis, K.A., & Chrysayi-Tokousbalides, M., (2006) On the mode of action of the phytotoxin (8R,16R)-(-)-pyrenophorin. *Pest. Biochem. Physiol., 86*,7-14.

Aliferis, K.A., & Chrysayi-Tokousbalides, M., (2006). Metabonomic Strategy for the Investigation of the Mode of Action of the Phytotoxin (5*S*,8*R*,13*S*,16*R*)-(-)-Pyrenophorol Using ^{1}H NMR Fingerprinting. *J. Agric. Food Chem., 54*, 1687-1692.

Aliferis, K.A., Chrysayi-Tokousbalides, M., & Fasseas, C. (2006). Physiological and ultrastructural changes in "green islands" on *Avena sterilis* leaves caused by (8*R*, 16*R*)-(-)pyrenophorin. *Plant Phys. Biochem., 44*, 851-856.

Andolfi, A., Boari, A., Evidente, A., & Vurro, M. (2005). Metabolites inhibiting germination of *Orobanche ramosa* seeds produced by *Myrothecium verrucaria* and *Fusarium compactum*. *J. Agric. Food Chem., 53*, 1598-1603.

Au, T. K., Chick, W. S. H. & Leung, P. C. (2000). Initial kinetics of the inactivation of calmodulin by the fungal phytotoxin ophiobolin A. *Int. J. Biochem. Cell Biol. 32,* 1173-1182.

Bakuzis, P., Bakuzis, M. L. F., & Weingartner, T. F. (1978). A new *β*-acyl vinyl anion equivalent. Synthesis of pyrenophorin. *Tetrahedron Lett.,19*, 2371-2374.

Baldwin, J. E., Adlington, R. M., & Ramcharitar, S. H. (1992). A short asymmetric synthesis of (*R*,*R*)-(-)-pyrenophorin. *Synlett., 11*, 875-877.

Bobylev, M. M., Bobyleva, L. I., & Strobel, G. A. (1996). Synthesis and bioactivity of analogs of maculosin, a host-specific phytotoxin produced by *Alternaria alternata* on spotted knapweed (*Centaurea maculosa*). *J. Agric. Food Chem., 44*, 3960-3964.

Bronson, C. R. (1991). The genetics of phytotoxin production by plant pathogenic fungi. *Experientia, 47*,771-776.

Bunkers, G. J., & Strobel, G. A. (1991). A proposed mode of action for green island induction by the eremophilane phytotoxins produced by *Drechslera gigantea, Physiol. Mol. Plant Pathol. 38*, 313-323.

Bunkers, G., Kenfield, D., Strobel, G., & Sugawara, F. (1990). Structure-activity relationships of the eremophilanes produced by *Drechslera gigantea. Phytochemistry, 29*, 1471-1474.

Capasso, R., Evidente, A., Cutignano, A., Vurro, M., Zonno, M. C., & Bottalico, A. (1996). Fusaric and 9,10-dehydrofusaric acid and their methyl esters from *Fusarium nygamai. Phytochemistry, 41*, 1035-1039.

Chen, S., Xu, X., Dai, X., Yang, C. & Qiang, S. (2007). Identification of tenuazonic acid as a novel type of natural phtotsystem II inhibitror binding in QB-site of *Chlamydomonas reinhardtii. Bioph. Bioch. Acta, 1767*, 306-318.

Chrysayi-Tokousbalides, M., (1997). Phoma metabolites toxic to *Convolvulus* spp. *Phytopath. Medit.,36*, 19-23.

Chrysayi-Tokousbalides, M., Machera, K., Kyriakopoulou, K., Aliferis, K. A., Schrader, K. K., Tsoutsanis, I., & Anastasiadou, P. (2007). Comparative toxicity of the phytotoxins (8*R*,16*R*)-(-)-pyrenophorin and (5*S*,8*R*,13*S*,16*R*)-(-)-pyrenophorol on aquatic organisms. *Bull. Envir. Contam. Toxicol.* (in press).

Colvin, E. W., Purcell, T. A., & Raphael, R. A. (1976). Total synthesis of the macrocyclic antibiotic (+/-)-pyrenophorin, *J. Chem. Soc., 16*, 1718-1722.

Dayan, F. E., Rimando, A. M., Tellez, M. R., Scheffler, B. E., Roy, T., Abbas, H. K., & Duke, S. O. (2002). Bioactivation of the fungal phytotoxin 2,5-anhydro-D-glucitol by glycolytic enzymes is an essential component of its mechanism of action. *Z. Naturforsh., 57c*, 645-653.

Dayan, F. E., Romagni, J. G., & Duke, S. O. (2000). Investigating the mode of action of natural phytotoxins. *J. Chem. Ecol., 26* (9), 2079-2094.

Dayan, F., Romagni, J., Tellez, M., Rimando, A., & Duke, S. (1999). Managing weeds with natural products. *Pestic. Outlook*, 185-188.

Desjardins, A. E., McCormick, S. P., & Appell, M. (2007). Structure-activity relationships of trichothecene toxins in an *Arabidopsis thaliana* leaf assay. *J. Agric. Food Chem.,*(on web).

Dommerholt, F.J., Thijs, L., & Zwanenburg B., (1991). Total synthesis of the macrodiolide pyrenophorol. Tetrahedron Lett. , 32, 1499-1502.

Duke, S. O., Dayan, F. E., Rimando, A. M., Schrader, K. K. Aliotta, G., Oliva, A., Romagni, J.G. (2002). Chemicals from nature for weed management. *Weed Sci. 50*, 138-151.

Duke, S. O., Romagni, J. G., & Dayan, F. E. (2000). Natural products as sources for new mechanisms of herbicidal action, *Crop Prot. 19*, 583-589.

Dumont, W., Vermeyen, C., & Krief, A. (1984). Novel synthetic route gamma-oxo-acrylates application to the synthesis of pyrenophorin antibiotic. *Tetrahedron Lett. 25*, 2883–2886.

Evidente, A., Andolfi, A., Cimmino, A., Vurro, M., Fracchiolla, M. & Charudattan, R. (2006). Herbicidal potential of ophiobolins produced by *Drechslera gigantea. J. Agric. Food Chem., 54*, 1779-1783.

Evidente, A., Andolfi, A., Vurro, M., Fracchiolla, M., Zonno, M. C., & Motta A. (2005). Drazepinone, a trisubstituted tetrahydronaphthofuroazepinone with herbicidal activity produced by *Drechslera siccans*. *Phytochemistry*, *66*, 715-721.

Evidente, A., Fiore, M., Bruno, G., Sparapano, L., & Motta, A. (2006). Chemical and biological characterisation of sapinopyridione, a phytotoxic 3,3,6-trisubstituted-2,4-pyridione produced by *Sphaeropsis sapinea*, a toxigenic pathogen of native and exotic conifers, and its derivatives. *Phytotchemistry, 67*, 1019-1028.

Evidente, A., Sparapano, L., Fierro, O., Bruno, G., & Motta, A. (1999). Sapinofuranones A and B, two new 2(3H)-dihydrofuranones produced by *Sphaeropsis sapinea*, a common pathogen of conifers. *J. Nat. Prod. 62*, 253-256.

Evidente, A., Sparapano, L., Motta, A., Giordano, F., Fierro, O., & Frisullo, S. (1996). A phytotoxic pimarane diterpene of *Sphaeropsis sapinea* f. sp. *cupressi*, the pathogen of a canker disease of cypress. *Phytochemistry, 42*, 1541-1546.

Gamboa-Angulo, M. M., Escalante-Erosa, F., Garcia-Sosa, K., Alejos-Gonzalez, F., Delgado-Lamas, G. & Pena-Rodriguez, L. M. (2002). Natural zinniol derivatives from *Altrenaria tagetica*. Isolation, synthesis, and structure-activity correlation. *J. Agric. Food Chem., 50*, 1053-1058.

Grove, J.F., (1971). Metabolic products of *Stemphylium radicinum*. IV. Minor products, *J. Chem. Soc., 12*, 2261-2263.

Harrington P. M., Singh, B. K., Szamosi, I. T., & Birk, J. H. (1995). Synthesis and herbicidal activity of cyperin. *J. Agric. Food Chem., 43*, 804-808.

Henkel, T., Brunne, R. M., Müller, H. & Reichel, F. (1999). Statistical investigation into the structural complementarity of natural products and synthetic compounds. *Angew. Chem. Inter. Ed., 38,* (5), 643-647.

Hoagland, R. E., (1990). Microbes and Microbial products as herbicides: An overview, in Microbes and Microbial products as herbicides, ED by Hoagland, R. E., ACS SymposiumSeries 439, American Chemical Society, Washington , DC, pp2-52.

Hoagland, R. E., (1991). Microbial allelochemicals and pathogens as bioherbicidal agents. *Weed Technol., 15*, 835-857.

Hoffman, R. V., Patonay, T., Nayyar, N. K., & Tao, J. (1996). A short, enantiospecific synthesis of a protected seco acid precursor to (*R,R*)-(-)-pyrenophorin. *Tetrahedron Lett. 37*, 2381-2384.

International Organization for Standardization, (2003). Water quality-Determination of toxic effect of water constituents and waste water to duckweed (*Lemna minor*)-Duckweed growth inhibition test. ISO/DIS 20079 (draft). Geneva, Switzerland.

Ishibashi, K., (1961), Studies on antibiotics from *Helminthosporium* sp. Fungi. Part II. Pyrenophorin, a new antibiotic produced by *Pyrenophora avenae* (=*Helminthosporium avenae*), *J. Agric. Chem. Soc. Japan, 35*, 257-262.

Kastanias, M. A. & Chrysayi-Tokousbalides, M., (1999). Comparative phytotoxicity of pyrenophorin and pyrenophorol isolated from a *Drechslera avenae* pathotype. *Proceedings of the 1999 Brighton Conference: Weeds,* 581-582.

Kastanias, M. A., & Chrysayi-Tokousbalides, M., (2000). Herbicidal potential of pyrenophorol isolated from a *Drechslera avenae* pathotype. *Pest Manag. Sci., 56*, 227-232.

Kastanias, M. A., & Chrysayi-Tokousbalides, M., (2005). Bioactivity of the fungal metabolite (8*R*,16*R*)-(-)-pyrenophorin on graminaceous plants. *J. Agric. Food Chem.,53*,5943-5947.

Kind, R., Zeeck, A., Grabley, S., Thiericke, R, & Zerlin, M. (1996). Secondary metabolites by chemical screening. 30. Helmidiol, a new macrodiolide from *Alternaria alternata. J. Nat. Prod., 59*, 539-540.

Kis, Z., Furger, P., & Sigg, H. P. (1969). Űber die Isolierung von Pyrenophorol. *Experientia, 25*, 123-124

Kobayashi, Y., Kishihara, K., & Watatani, K. (1996).*Two-step conversion of 2-substituted furans into* γ-oxo-α,β-unsaturated carboxylic acids. Formal synthesis of (+)-patulolide A and (-)-pyrenophorin. *Tetrahedron Lett., 37*, 4385-4388.

Lerario, P., & Graniti, A. (1985). Attivitá fitotossica della pyrenoforina e sua produzione nelle colture di *Pyrenophora avenae* Ito et Kurib. *Phytopath. Medit., 24*, 280-283.

Machinaga, N., & Kibayashi, C. (1993). Preparation of macrodiolides via a common chiral building block. Total synthesis of (-)-pyrenophorin and (-)-pyrenophorol. *Tetrahedron Lett., 34*, 841-844.

Michel, A., Johnson, R. D., Duke, S. O. & Scheffler, B. E. (2004). Dose-response relationships between herbicides with different modes of action and growth of *Lemna paucicostata*: an improved ecotoxicological method. *Environ. Toxicol. Chem. 23*, 1074-1079.

Ngooi, T. K., Scilimati, A., Guo, Z. W., & Sih, C. J., (1989). A chemoenzymatic route to (-)-pyrenophorin. *J. Org. Chem., 54*, 911.

Nokami, J., Takeuchi, T., Gomyo, S., & Kakihara, T. (1994). A new approach to the synthesis of γ-hydroxy-α,β-unsaturated macrolides and (-)-pyrenophorin by intramolecular C=C bond formation with oxidative functionalization from ω-[α-(*p*-chlorophenylsulfinyl)acetoxy]alkanal. *Chem. Lett. 23*, 1103-1106.

Pedras, M. S. C., & Biesenthal, C. J. (2001). Isolation, structure determination, and phytotoxicity of anusual dioxopiperazines from the phytopathogenic fungus *Phoma lingam*. *Phytochemistry, 58*, 905-909.

Pedras, M. S. C., & Chumala, P. B. (2005). Phomapyrones from blackleg causing phytopathogenic fungi: isolation, structure determination. Biosyntheses and biological activity. *Phytochemistry, 66*, 81-87.

Pedras, M. S. C., Biesenthal, C. J., & Zaharia, I. L. (2000). Comparison of the phytotoxic activity of the phytotoxin destruxin B and four natural analogs. *Plant Sci., 156*, 185-192.

Pedras, M. S. C., Chumala, P. B., & Quail, J. W. (2004). Chemical mediators: The remarkable structure and host-specificity of despilairdin, a sesquiterpenic depsipeptide containing a new amino acid. *Org. Lett., 6* (24), 4615-4617.

Pedras, M. S. C., Taylor, J. L., & Nakashima, T. T. (1993). A novel chemical signal from the "blackleg" fungus: Beyond phytotoxins and phytoalexins. *J. Org. Chem, 58*, 4778-4780.

Pena-Rogriguez, L.M., Armigeon, N. A., & Chilton, W. S. (1988). Toxins from weed pathogens, I. Phytotoxins from a *Bipolaris* pathogen of Johnson grass. *J. Nat. Prod., 51*, 821-828.

Rendon-von Osten, J., Ortiz-Arana, A., Guilhermino, L., Soares, A. M. V. M. (2005). *In vivo* evaluation of three biomarkers in the mosquitofish (*Gambusia yucatana*) exposed to pesticides. *Chemosphere, 58*, 627-636.

Ruiz, M. J., Lopez-Jaramillo, M., Redondo, G., Font, G. (1997). Toxicity assessment of pesticides using the Microtox test: Application to environmental samples. *Bull. Environ. Contam. Toxicol. 59*, 619-625.

Sakamura, S., Ichihara, A., & Yoshihara, T. (1988).Toxins of phytopathogenic microorgasnisms: Structural diversity and physiologicl activity, in *Biologically Active Natural Products: Potential use in Agriculure*, Ed by Cutler H G., ACS SymposiumSeries 380, American Chemical Society, Washington , DC, pp57-64.

Schrader, K. K.; Rimando, A. M.; & Duke, S. O. (2002). *Studies in natural products chemistry*. Atta-ur-Rahman (ed), 26, Amsterdam, The Netherlands: Elsevier.

Seuring, B., & Seebach, D. (1978). Synthese und Bestimmung der absoluten Konfiguration von Norpyrenophorin, Pyrenophorin und Vermiculin. *Liebigs Ann. Chem., 12*,2044-2073.

Solfrizzo, M., Vitti, C., De Girolamo, A., Visconti, A., Logrieco, A., & Fanizzi, F. P. (2004). Radicinols and radicidin phytotoxins produced by *Alternaria radicina* on carrots. *J. Agric. Food Chem., 52*, 3655-3660.

Somasundaram, L., Coats, J. R., Racke, K. D., & Stahr, H. M. (1990). Application of the Microtox System to assess the toxicity of pesticides and their hydrolysis metabolites. *Bull. Environ. Contam. Toxicol. 44*, 254-259.

Sugawara, F., & Strobel, G. A., (1986). (-)-Dihydropyrenophorin, a novel and selective phytotoxin produced by *Drechslera avenae*. *Plant Sci., 43*, 1-5

U. S. Environmental Protection Agency, (1996). Ecological effects test guidelines OPPTS 850.4400. Aquatic plant toxicity test using *Lemna* spp., Tiers I and II. EPA 712-C-96-156. Washington DC, p 10.

Vassiliou, G., Chrysayi-Tokousbalides, M., Charitoudis, S., & Aliferis, K. A. (2005). Sensitivity of *Gambusia affinis* to the fungal metabolite (8*R*,16*R*)-(-)-pyrenophorin. *Proceedings of 12th International Symposium on Toxicity assessment*, 149.

Venkatasubbaiah, P., & Chilton, W. S. (1992). An epoxydon-derived ester from a *Phoma* sp. pathogenic to rhubarb. *J. Nat. Prod., 55* (5), 639-643.

Vesonder, R. F., Labeda, D. P., & Peterson, R. E. (1992). Phytotoxic activity of selected water-soluble metabolites of *Fusarium* against *Lemna minor* L. (Duckweed). *Mycopathologia, 118*, 185-189.

Vischetti, C. & Esposito, A. (1999). Degradation and transformation of a potential natural herbicide in three soils. *J. Agric. Food Chem., 47*, 3901-3904.

Walton, J. D., & Panaccione, D. G., (1993). Host-selective toxins and disease specificity: Perspectives and progress. *Ann. Rev. Phytopath, 31*, 275-303.

Ward, D. E., Vazquez, A., & Pedras, M. S. C. (1996). Understanding host-selective phytotoxicity: Synthesis and biological discrimination of phomalide and its (*Z*)-isomer. *J. Org. Chem, 61*, 8008-8009.

In: Weeds: Management, Economic Impacts and Biology
Editor: Rudolph V. Kingely
ISBN: 978-1-60741-010-2

Short Communications

COST BENEFIT IN CABBAGE CULTIVATION BY WEED CONTROL USING ORYZALIN HERBICIDE AND HOEING TECHNOLOGY

Md. Wasim Aktar[1]*, Dwaipayan Sengupta[2] and Samsul Alam[3]

[1]Pesticide Residue Laboratory, Department of Agricultural Chemicals,
Bidhan Chandra Krishi Viswavidyalaya, Mohanpur-741252,
Nadia, West Bengal, India
[2]Department of Agricultural Chemistry and Soil Science,
Institute of Agricultural Science, University of Calcutta,
Kolkata, West Bengal, India
[3]Institute of Pesticide Formulation Technology
Sector 20, Udyog Vihar, Opp. Ambience Mall (on NH-8), Gurgaon-122016,
Haryana, India

ABSTRACT

One pre-emergent application of oryzalin 40% SC herbicide and one hoeing was found to control maximum pests of cabbage crop. By following proper date of sowing of cabbage crop and controlling the weeds, no pest including insect and diseases was observed to attack the crop. During the period of controlling weeds of cabbage crop neither insect pest nor any fungal attack was observed during the period of this study. The insects and diseases were thus below economic threshold value.

Keywords: cabbage cultivation, oryzalin, herbicide, economics analysis.

---

* Correspondence to: wasim04101981@yahoo.co.in

1. Introduction

Cabbage *(Brassica oleracea* L. var *capitata) is* one of the important winter vegetables grown in India. It is attacked by number of pests including weeds. As the losses caused by insect pests and diseases of cabbage crop are visible, various studies have been reported about their effective control (Geeson *et al.*,1979; Bedlan, 1998; Panda *et al,*1999; Arora *et al,*2005 and Shi Zu Hua *et al,* 1998). Since the weeds compete for nutrients with the plant and act as host for disease causing pathogens, they cause indirect loss (Arora *et al,* 1998). The weed control in this crop has not been studied that extensively. Major weeds of this crop are *Coronopus didymus* and *Melilotus indica* and minor weeds are *Partheniun hysterophorus, Rumex dentatus, etc* in and Nadia district of west bengal. The major pests of cabbage are Diamond Back Moth (DBM) and *Alternaria solani.* It was envisaged that early sowing could enable early harvesting before the onslaught of insect (DBM) attack. Adverse affect of Alternaria diseases[7] could be avoided in cold winter months (Arora *et al,* 2006). The study was conducted to validate the proposed package.

2. Materials and Methods

2.1. Design of Experiment

Cabbage (var. Rareball) was raised in the fields of Mandauri at B.C.K.V Mohanpur, Nadia, West Bengal, India using randomized block design (RBD). The plot size was 5 X 2 m^2 with 40 cm distance between plants and rows. The date of sowing was 23rd October 2006. During the study average temperature was maximum 27.6^{0}C and minimum 12.2^{0}C. The average conditions were temperature 17.9^{0}C, relative humidity 59.5%, and sunshine 6.6 h. There was small rainfall (Average 16 mm.) during the period.

2.2. Collection of Sample

The treatments were i) T_1 - one hoeing, ii) T_2 oryzalin @ 1.5 kg ai ha^{-1}, iii) T_3 - oryzalin @ 3 kg ai ha^{-1}, iv) T_4 - oryzalin @ 1.5 kg ai ha^{-1} + one hoeing, v) T_5 - oryzalin @ 3 kg ai ha^{-1} + one hoeing and vi) T_6- control. The herbicide was applied 10 days before transplanting of seedlings of cabbage while hoeing was carried out after 20 days of transplanting. The insect pest DBM count was observed on 4 plants from each sub plot. The population of various weeds was counted in plots using a quartet, 1 m^2, before harvesting the crop. Fresh weight of individual weeds from each plot was taken.

3. Results and Discussion

The population of major and minor weeds is presented in Table 1. The data for treatment T_6 (control) under fresh biomass revealed that among the weeds, *Coronopus didymus* was the major weed with highest density 56.5 weeds per m^2 and bio-mass 136.4 g m^{-2} followed by

Phalaris minor under bio-mass 52.6 g m^{-2}. Under weed density, it was followed by *Melilotus indica,* which counted 41 weeds per m^2.

After giving one hoeing without the application of oryzalin, the weed count of C. didymus was reduced to 121.4 in T_1, from 136.4 in T_6. Application of oryzalin at normal dose (T_2) reduced the population *C. didymus to* 11.5 weeds per m^2 and to 0.5 weeds per m^2 with double dose (T_3). While *M. indica* was completely controlled, P. *minor* density was reduced to 2.0 in treatment T_2, Similar results were observed with biomass of weeds in treated plots. One application of oryzalin along with one hoeing further reduced the *C. didymus* population to 28.5 (T_4).

Double dose of oryzalin along with one hoeing (T_5) could reduce the weed almost completely. But since double dose resulted in phytotoxicity to the cabbage and led to yield reduction (Table 2), treatment T_4, involving one application of oryzalin *at* the recommended rate of application followed by hoeing was found to optimum.

The fresh as well as dry weight of weeds was reduced, as the number of weeds was reduced. No other pest like DBM, *Helicoverpa, alternaria* or *xanthomonas* was observed above their economic threshold level except weeds, so only weed population count was undertaken. The yield of cabbage from the plots treated with oryzalin was higher as compared to that obtained from control (untreated) plots. The yield was still higher in the treated plots, where besides oryzalin treatment; one additional hoeing was carried out (Table 2). The statistical analysis of yield data in Table 2 shows that CD calculated at 5% level is 3.624 q ha^{-1} and at 1 % level is 5.155 q ha^{-1}. The CD at both 1 % and 5% levels is quite lower than the difference of T_6, T_1; T_2, T_4 ; and T_4 , T_3 which indicates that treatments T_6, T_1, T_2, T_4 ; and T_4 , T_3, are significantly different from each other but not T_3, and T_5,. The data in Table 1 and 2 clearly indicates that the herbicide is highly effective for weed control of cabbage crop. However results displayed in Table 2 show that on applying double the recommended dose yield was lower than that realized on applying the herbicide at recommended dose and therefore should not be recommended.

The higher yield of cabbage was realized from the plots treated with one application of recommended dose of oryzalin along with one hoeing, compared to the treated plots without hoeing and untreated as well as treated plots receiving higher than recommended dose. The treatment T_6 with highest number of weeds and therefore highest total bio-mass resulted in reduction of cabbage yield. T_6 treatment therefore gave least yield of cabbage crop. But treatment T_3 and T_5, showing least number of weeds and bio-mass also recorded less yield. The reason being the phytotoxicity caused by the herbicide to the cabbage crop. Applying double the recommended dose causes more expense and yield of the produce is also lowered. Considering the selling price of cabbage as Rs 75 q^{-1} in Nadia, W.B, the additional earnings can be calculated as Rs 22,000 ha^{-1}.

Oryzalin along with hoeing controlled weeds (Table 1). Date of sowing helped in avoiding onslaught of insect and disease attack. As the inoculums of disease and insect harbor in weeds, it was envisaged that control of weeds and sanitation in field due to herbicide application and one hoeing led to improved yield as shown in Table 2. It can be therefore concluded that if date of sowing of cabbage crop is delayed for 10-15 days, there is a definite

Table 1. Bio mass and density of weeds under different treatments

Weeds	Fresh bio mass (gm^{-2})					Density (Numbers m^2)					
	T_1	T_2	T_3	T_4	T_6	T_1	T_2	T_3	T_4	T_5	T_6
Coronopus didymus	121.4 (18.4)	31.3 (7.2)	2.1 (0.19)	28.5 (6.8)	136.4 (21.0)	49.5	11.5	0.5	9.5	0.0	56.5
Melilotus indica	6.9 (0.8)	4.6 (1.1)	-	4.0 (0.92)	5.1 (0.9)	39	2.5	-	2.2	-	41.0
Phalaris minor	49.6 (7.1)	45.1 (10.4)	7.8 (1.2)	40.7 (0.98)	52.6 (8.0)	6.8	3.5	2.0	1.8	0.9	3.5
Canabis sativa	7.6 (1.2)	0.2 (0.04)	-	0.0	10.4 (1.5)	3.1	0.06	-	0.0	-	4.2
Rumex dentatus	20.4 (2.9)	3.1 (0.5)	1.6 (0.1)	2.8 (0.3)	22.6 (3.4)	7.2	0.9	0.3	0.5	0.0	8.2
Anagallis arvensis	0.96 (0.1)	-	-	-	1.3 (0.2)	-	-	-	-	-	2.5
Total	206.86	84.3	11.5	76.0	228.4	106.5	18.46	2.8	14.0	0.9	120.9

Figures in parenthesis indicate dry weight; Treatments : T1 - one hoeing, T2- 1.5 Kg ha- 1 oryzalin, T 3-3 Kg ha-1 oryzalin, T4 – 1.5 Kg ha-1 oryzalin + hoeing, T5 - 3 Kg ha- I oryzalin + hoeing, T 6 - Control (No treatment).

Table 2. Economics of cabbage cultivation following hoeing and pendimethalin treatments

Treatments	Yield (q ha-1)	Production cost (Rs)	Gross income (Rs)	Net Returns (Rs)
T_1	439.3	13,740	32,947	19,207
T_2	433.7	13,430	32,527	19,097
T_3	387.0	14,410	29,025	14,615
T_4	488.3	14,630	36,622	21,992
T_5	407.3	15,120	30,547	15,427
T_6	330.0	12,540	24,750	12,210

$CD_{5\%}$ = 3.624; $CD_{1\%}$ =5.155; Treatments : T_1 - one hoeing, T_2 – 1.5 Kg ha^{-1} pendimethalin, T_3 - 3 Kg ha^{-1} pendimethalin, T_4 – 1.5 Kg ha^{-1} pendimethalin + hoeing, T_5- 3 Kg ha^{-1} pendimethalin + hoeing, T_6 - Control (No treatment); 1 US$= 42.00 Indian Rupees (Rs.) approx.

pest attack, especially DBM on cabbage crop even after following weed control practices. Oryzalin at recommended dose of application in addition with one hoeing (necessary practice after transplanting of crop) and proper date of sowing of cabbage crop gave excellent yield results among other treatments applied in cabbage crop.

CONCLUSION

The use of a pre-plant application of oryzalin along with hoeing and right time of transplanting can be recommended for cabbage cultivation. Oryzalin was found to be a persistent herbicide (Blanco *et al*,1988 and Speich,2003). Moreover by controlling weeds of cabbage crop neither insect pest nor any fungal attack was observed during the period of this study. The insects and diseases were thus below economic threshold value. The study confirms the report by Saimbhi *et al*,1992 in which oryzalin used for weed control gave disease control in cabbage.

REFERENCES

Geeson JD and Browne KM *(1979)* Effect of post harvest fungicide drenches on stored winter white cabbage. *Plant Pathol. 28: 161-168.*

Bedlan G *(1998)* Application management to control post harvest Botrytis infections on Chinese cabbage. *Gesunde Pflanzen 50: 162-164.*

Panda SK, Behera UK and Nayak SK *(1999)* Bio-eff icacy of fipronil 5% SC against diamond back moth, *Plutella xylostella* (Linn.) (Yponomeutidae, Lepidoptera) infesting cabbage. *Insect Environment 5: 104-105.*

Arora S and Gopal M *(2005)* Potential of fipronil and *Bacillus thuringiensis* to control *Plutella xylostella* on cabbage crop. *International Pest Control 47 (5): 251-256.*

Shi Zu Hua, Shu Sheng, Liu, Shi ZH and Liu SS *(1998)* Toxicity of insecticides commonly used in vegetable fields to the diamondback moth, *Plutella xylostella* and its parasite, *Cotesia plutellae. Chinese Journal of Biological Control 14: 53-57.*

Arora S and Mittal A *(1998)* Integrated weed management. *Indian farming. 48:* 20.

Arora S and Gopal, M *(2006)* Bioeff icacy of Iprodione against two desapers, its compatibility with T. *harzianum* and residues on cabbage crop. *J Environ Sci health IS 41 (6): 949-963.*

Blanco,H.G., Oliveira,D.A. and Matallo,M.B. (1988) Persistence and residues of oryzalin in soils under soybean crop.*Pesquisa Agropecuaria Brasileira*. 23: 10, 1107-1113.

Speich, P. (2003) Chemical weed control: modifications of practices and their impact on transfer of chemicals to water.*Progres Agricole et Viticole*. 120: 5, 113-118.

Saimbhi MS, Sandhu KS, Singh D, Kooner KS, Dhiman JS, Dhillon INIPS and Singh DU *(1992).* Use of herbicide for weed and disease control in cabbages. *Advances in Horticultural Science 6: 125-128.*

In: Weeds: Management, Economic Impacts and Biology
Editor: Rudolph V. Kingely
ISBN: 978-1-60741-010-2

Short Communications

HERBICIDES-THEIR BIOLOGICAL ACTIVITY, TRANSFORMATION, METABOLISM AND ENVIRONMENTAL IMPACT UNDER INDIAN SUBCONTINENT

A. Chowdhury[1*], M. K.Bhowmick[2], P.K. Biswas[3], R. B. Mallick[1] and Saswati Pradhan[1]

[1] Institute of Agricultural Science, University of Calcutta, Kolkata-700 019 (W.B.), India
[2]Rice Research Station (Govt. of W.B.), Chinsurah (R.S.)-712 102 (W.B.), India
[3]Soil Testing Laboratory (Govt. of W.B.), Raiganj-733 130 (W.B.), India

1. INTRODUCTION

Since man first began to cultivate crops, undesirable plants called weeds have been a problem. Weeds have long been recognized as major pests causing significant losses in crop productivity (Bhowmick and Ghosh, 2002a; Bhowmick and Biswas, 2007), because, it is unfortunate that the damage on crops by weeds is invisible as compared to other pests. Farmers appreciate weeds as a problem only when it goes out of hand, or when they seriously affect agricultural operations such as harvesting (e.g. spiny weeds). In addition to direct effect on crop yield, weeds result in considerable reduction in the efficiency of inputs used. The precious and costly inputs such as fertilizers and irrigation water, which are otherwise meant for realizing the potential yield, are depleted by the weeds (Yaduraju and Mishra, 2005; Bhowmick and Biswas, 2007).

India with 2.4% of world's geographical area, is sustaining about 17.7% of the global human population (Ghosh and Bhowmick, 2008). Between 1950-51 and 2006-07 production of foodgrains increased at an annual rate of 2.5% compared to the average population growth rate of 2.1%, due to which India remained self-sufficient in foodgrains till 2005-06. The scenario, however, changed between 1990 and 2007 when the growth rate of foodgrain

* Correspondence: Prof. Ashim Chowdhury, Acharya P. C. Ray Professor in Agricultural Chemistry, E mail- ashimchowdhry@gmail.com ; ashimkly@hotmail.com

production decelerated to 1.2% which was lower than the average population growth rate of 1.9% (Ramadhar, 2008). It was estimated that to feed around one billion Indian population, in this century, about 225 million tons of foodgrains would be required. This is a challenge ahead of us. New crop production technology must cope with this challenge, in relation to inputs of farm productivity. Weed management is one of the significant inputs, especially in less industrialized (developing) countries (Bhowmik, 1999). Indian agriculture is, thus, facing an arduous task to provide food security as well as nutritional security for all, which will, hopefully, be possible by adopting the frontier technologies in every sphere of weed science.

Though chemical herbicides represent the most effective tool to combat the weed menace, their indiscriminate or injudicious use has led to concerns about herbicide resistance, residue and toxicity problems, weed shift and resurgence, and environmental pollution. This has warranted either a shift in alternative weed management strategy from the use of chemical herbicides to that of bioherbicides i.e. the use of plant pathogenic organisms *viz.* fungi, bacteria and viruses (Bhowmick and Ghosh, 2002a) or a judicious use of chemical herbicides along with utilization of all other available techniques/ approaches in a compatible and integrated manner. The present chapter is intended to illustrate the aspects of weed problems, benefits of and strategies for weed management, crop-wise weed management options, environmental impact of some chemical herbicides as well as bioherbicides and research priorities with regard to eco-friendly weed management.

2. Economic Impact of Weeds

Weeds are a component of any crop production system. The competitive effects that weeds have on the production of numerous crops are well-documented in most of the countries in the world. Herbicides have become essential for economically controlling the weeds in various crop production systems (Bhowmik, 1999).

In general, impact of weeds on crop yields varies from high input to low input crop production systems. It has been estimated that weeds caused 5% loss in agricultural production in most developed countries, 10% loss in developed countries and 25% loss in least developing countries (Bhan and Sushilkumar, 1998; Bhowmik, 1998 and 1999).

The world-wide loss of attainable production of eight principal food and cash crops (rice, wheat, barley, maize, potato, soybean, cotton and coffee) due to weeds has been estimated (between 1988 and 1990) to be $ 76.3 billion out of total loss of $ 243.7 billion caused due to pests annually (Oerke *et al.*, 1994). These losses are not shared equitably and developing countries shared the burden of greater losses than the industrialized world (Bhowmik, 1999). According to Parker and Fryer (1975), global food loss is reported to be 287 million tons due to weeds accounting to 11.5% of the total food production

Recent estimates showed that the average annual yield losses due to various crop pests in India were valued at about 60,000 crores of rupees (Singh, 2005), out of which weeds alone accounted for the maximum loss (Table 1). Estimates also showed that average crop yield reductions due to weeds vary from 29 to 72%, depending on the crops (Table 2). However, weeds in general reduce crop yields by 31.5% (22.7% in winter and 36.5% in summer and rainy seasons). But as farmers adopt some kind of weeding practices on their fields, a conservative estimate of 10% loss in crop yields may be taken as more realistic (Bhan and

Sushilkumar, 1998). Besides reduction in yields, weeds remove 30-40, 10-15 and 20-40 kg ha^{-1} of nitrogen, phosphorus and potassium, respectively (Alagh, 1988).

Table 1. Annual losses caused by various pests in India[a]

Pests	Loss (%)	Value (Rs. in crores)
Weeds	33	19,800
Diseases (Pathogens)	26	15,600
Insects	26	15,600
Rodents and others	15	9,000
Total	**100**	**60,000**

[a] Bhowmick and Biswas (2007)

Table 2. Average reduction in yield of important crops due to weeds in India[b]

Crop	Yield reduction (%)	Crop	Yield reduction (%)
Direct seeded rice	64.0	Chickpea	48.1
Transplanted tice	43.9	Field pea	47.1
Wheat	31.6	Cotton	60.4
Sugarcane	34.4	Groundnut	63.6
Potato	29.0	Mustard	37.7
Maize	48.8	Sesame	54.2
Soybean	53.0	Linseed	62.7
Greengram	46.3	Sunflower	32.2
Blackgram	55.4	Sugarbeet	70.3
Pigeonpea	46.7	Carrot	72.5
Lentil	58.8	Onion	68.0

[b] Modified after Bhan (1997), and Yaduraju and Mishra (2005)

In addition to direct reduction in crop yield through competition for soil moisture, nutrients, sunlight and space, weeds indirectly limit production by serving as alternate hosts for various pest organisms that adversely affect crop plants (Bhowmick, 2002). Admixture of weed seeds such as *Argemone mexicana* in mustard (Bhowmick, 2001), wild rice in rice, wild oat in wheat and wild onion in onion impairs the quality of crop produce. Various aquatic weeds like water hyacinth, *Salvinia* etc. impede flow of water in irrigation and drainage channels, canals and streams, and even cause flooding and sedimentation. Spiny or thorny weeds like *Argemone mexicana*, *Amaranthus spinosus*, etc. restrict free movement of farm workers in carrying out the farm practices. Some weeds like *Parthenium hysterophorous* L. (Wild carrot weed), *Lathyrus sativa* L., *Ambrosia artemisiifolia* L. (ragweed) and *Rhus radicans* L. (poison ivy) etc. cause itching, hay fever, illness and suffering to labourers when they come in contact. Parasitic weeds like *Cuscuta*, *Orobanche*, *Striga*, etc. are also posing a serious threat to field crops (Bhowmick and Mandal, 2001; Yaduraju and Mishra, 2005).

Thus, weeds are responsible, both directly and indirectly, for the colossal losses of food meant for human being (Bhowmick and Mandal, 2001).

3. Benefits of Weed Management in Agriculture

Weeds compete with desirable plants, and some unwanted plants are a threat to animal health and productivity. Weed management is an important part of the overall land management plan. Good management follows the saying: an ounce of prevention is worth a pound of cure. However, some of the benefits are listed hereunder:

- Buildup of damaging or toxic (poisonous) weed populations and their seeds gets prevented.
- Yield potential of a crop variety at seeding time is 100% and to maintain this potential, weed control must be imposed before any of the growth limiting factors *viz.* nutrients, sunlight, water and space becomes operating Thus, managing the weeds helps to maintain a desirable crop growth and yield potential.
- There is an increased pasture quality, forage yield, animal capacity, and animal health protection.
- Weed management raises the farm income through efficient energy conservation and labour utilization.

4. Weed Management Strategies

In India, till seventies, manual and mechanical removal was one of the best options available with the farmers to manage weeds in crop fields (Bhowmick and Ghosh, 2002a). During seventies and eighties, weed management was through the use of herbicides as the problems were specific and probably, herbicides provided answer because of their selective nature in the identical crop-weed morphological situation, eg. *Phalaris* in wheat and *Echinochloa* in rice. In the seventies, biological control also started to take shape for weed control and in the eighties good success was achieved to control water hyacinth (*Eichhornia crassipes* Solms) and water-fern (*Salvinia molesta* Mitchell). During nineties and onwards the scenario has been changing due to the various problems encountered in the herbicide use in different crops and cropping systems (Bhan and Sushilkumar, 1998).

The anticipated scenario which is now getting visible includes use of new chemical compounds. This may include use of botanicals, mycoherbicides and allelochemicals which can be used and get degraded automatically in the Mother Nature. The relative selectivity, utilization and development of these technologies are in progress and they may become a part of the integrated weed management (IWM) systems in the time to come. The other part of the scenario would be the change in the weed shift due to continuous use of herbicides and accordingly, strategies need to be developed for utilizing cropping systems as an approach and developing practices for changing the microclimate with the use of various practices in a particular cropping system (Bhan and Sushilkumar, 1998). However, there are various weed management strategies (discussed hereafter) which can be used with varying degree of success in different agro-ecological zones as well as production systems.

4.1. Preventive Approach in Weed Management

Preventive approach encompasses all measures taken to forestall the introduction and spread of new weeds (Bhan and Sushilkumar, 1998; Yaduraju and Mishra, 2005). Such measures include use of weed free crop seeds / seedlings for planting; use of thoroughly decomposed organic manures having no viable weed seeds; removal of weeds from irrigation ditches, fence rows, non-crop areas, etc.; using weed seed screens to filter irrigation water; restricting livestock movement into no-weed infested area; use of clean harvesters and tillage implements before moving to non-weed infested area; seed certification / weed laws / seed laws / quarantine laws; etc.

4.2. Physical Approach in Weed Management

Despite the great progress made in agriculture, manual and mechanical methods still continue to be important weed management practices in many regions of the world (Yaduraju and Mishra, 2005). But they have certain limitations such as unavailability of labours during peak period, high labour cost and unfavourable environment (Bhan and Sushilkumar, 1998).

4.3. Cultural Approach in Weed Management

Cultural methods are used to complement manual and mechanical methods. Cultural practices such as land preparation / tillage, timely weeding, adjusting planting / harvesting dates, cover crops, use of weed competitive crops and cultivars, proper seed rate / crop geometry and crop density, stale seed bed technique, crop rotation, intercropping, nutrient management, water management, mulching, soil solarization, etc. have pronounced effect on crop-weed interference. They are not only eco-friendly but also reduce the use of costly herbicides (Bhan and Sushilkumar, 1998; Yaduraju and Mishra, 2005).

4.4. Chemical Approach in Weed Management

Although cultural method of weed control is to be given topmost priority in recommending measures to solve the weed problems, yet farmers have developed inclination to use chemical method of weed control (Mukhopadhyay, 1992) as it is more labour and energy efficient than other methods. The progressive modernization of Indian agriculture involving intensive use of herbicides to control weeds is gaining importance in recent years. A fundamental reason for the widespread use of herbicides in modern agriculture is their ability to control selectively a wide range of weeds in a variety of crops and in some situations where all other methods fail or cannot be adopted (Mukhopadhyay, 1992). But the exclusive reliance on herbicide chemicals has led to concerns about contamination of environment by the presence of herbicide residue in soil, water and plants, particularly vegetables and fruits; threats to human health; toxicity to animals; shift in weed flora; and appearance of resistant weed species (Hatzios, 1987; Mukhopadhyay, 1992, 1993 and 1997; Gautam and Mishra, 1995; Bhowmick and Mandal, 2001; Bhowmick and Ghosh, 2002a). New approaches are, therefore, needed to manage weeds effectively while eliminating or averting such concerns.

Broadly, there are three categories of herbicides which can be used for effective chemical weed management in crop as well as non-crop areas:

a) Pre-plant incorporation (PPI) in soil: Herbicides having greater toxicity on the emerging crop seedlings and getting decomposed due to solar radiation are applied before the crop is planted. They are incorporated in the soil upto a depth of 5 cm after the final land preparation. e.g. fluchloralin, trifluralin, etc.

b) Pre-emergence (PE) application: Herbicides are applied after sowing but before the emergence of both the crops and weeds, preferably within 48 hours after sowing. e.g. pendimethalin, alachlor, butachlor, metolachlor, oxadiazon, linuron, etc.

c) Post-emergence (POE) application: Herbicides are applied after the emergence of crops or weeds, around 10-25 days after sowing (DAS). e.g. 2,4-D, isoproturon, bentazone, chlorimuron ethyl, sulfosulfuron, clodinafop, fluazifop-p-butyl, etc.

4.5. Biological Approach in Weed Management

Biological weed control is an approach that uses living organisms to control or reduce the population of a selected undesirable weed species, while leaving the crop unharmed (TeBeest, 1991). The concept of deliberately using organisms to control a pest constitutes the fundamental basis of all biological pest control systems (Hatzios, 1987). Sustainable agricultural systems dictate that inputs currently provided by nonrenewable petrochemical resources should be replaced by biologically based renewable inputs (Quimby *et al.*, 2002). Biological control of weeds using plant pathogens is accepted as a practical, safe, and environmentally beneficial weed management method applicable to agroecosystems (Chrudattan, 2001). Extensive research on this aspect has resulted in the establishment of three major strategies for the biocontrol of weeds, namely, the classical (inoculative), the bioherbicide (inundative), and the conservation and augmentation biological control (TeBeest, 1996; Ghorbani *et al.*, 2005; Yaduraju *et al.*, 2005).

In the classical strategy, a biocontrol agent is simply introduced or released into a weed population to establish itself and control the weed population, requiring no further manipulation (Templeton *et al.*, 1979; Charudattan, 1984). Insects, fungi, nematodes, fish, and other biological systems have all been used with varying degrees of success as biological agents in this strategy (Charudattan and Walker, 1982). Examples of successful classical approach of biological weed control in India are use of curculionid weevils *Neochetina bruchi* Hustache and *N. eichhorniae* Warner to control *Eichhornia crassipes* (Martius) Solms (water hyacinth) and use of weevil *Cyrtobagous salvinae* Calder and Sands to control *Salvinia molesta* D.S. Mitchell (waterfern/African payal); and use of grasscarp fish

(*Ctenopharyngodon idella* Vahl.) for controlling submerged aquatic weeds (Mukhopadhyay, 1986 and 1993; Singh, 2001; Yaduraju *et al.*, 2005).

The bioherbicide strategy employs the abundant/massive, usually annual, release of a biocontrol agent into specific weed-infested fields to infect and kill susceptible weeds (Templeton *et al.*, 1986; Yaduraju *et al.*, 2005). This has been discussed in detail later on.

Regarding the conservation and augmentation strategy, many authors define conservation biological control as actions that preserve or protect natural enemies and augmentation as actions that indirectly increase the populations of natural enemies. Other researchers consider conservation to mean environmental modification to protect and enhance natural enemies (Ghorbani *et at.*, 2005). Here, it is necessary to apply relatively low dose of inoculum to reach a high enough level of the agent organism required for an epidemic or outbreak. In this strategy, most of the efforts have focussed on use of indigenous pathogens or insects to control indigenous weeds (Yaduraju *et al.*, 2005). This is similar to the bioherbicide strategy in that while there is direct human manipulation and distribution of inoculum, the inoculum is neither mass-produced nor applied as an inundative dose over large areas (Charudattan, 1984).

4.6. Ecological Approach in Weed Management

Because of environmental and human health concerns, worldwide efforts are being made to reduce the heavy reliance on synthetic herbicides that are used to control weeds. In this regard, the phenomenon of allelopathy, which is expressed through the release of chemicals by a plant, has been suggested to be one of the possible alternatives for achieving sustainable weed management. The use of allelopathy for controlling weeds could be either through directly utilizing natural allelopathic interactions, particularly of crop plants, or by using allelochemicals as natural herbicides. In the former case, a number of crop plants with allelopathic potential can be used as cover, smother, and green manure crops for managing weeds by making desired manipulations in the cultural practices and cropping patterns. These can be suitably rotated or intercropped with main crops to manage the target weeds (including parasitic ones) selectively. Even the crop mulch/residues can also give desirable benefits. Not only can the terrestrial weeds, even allelopathy be suitably manipulated for the management of aquatic weeds. The allelochemicals present in the higher plants as well as in the microbes can be directly used for weed management. Their bioefficacy can be enhanced by structural changes or by synthesis of chemical analogues based on them. Further, in order to enhance the potential of allelopathic crops, several improvements can be made with the use of biotechnology or genomics and proteomics. In this context, either the production of allelochemicals can be enhanced or the transgenics with foreign genes encoding for a particular weed-suppressing allelochemical could be produced. In the former, both conventional breeding and molecular genetical techniques are useful. However, with conventional breeding being slow and difficult, more emphasis is laid on the use of modern techniques such as molecular markers and selection aided by them. Although the progress in this regard is slow, nevertheless some promising results are coming and more are expected in future. Allelopathy is a very useful technique, providing a new possibility for sustainable weed management. The status of research on allelopathy and future scope of work in India has been well reviewed by Bansal and Bhan (1993).

4.7. Biotechnological Approach in Weed Management

Application of biotechnology can play a great role to solve the emerging problems of chemical weed management by standard herbicide. At present, four major areas within the field of weed management offer attractive opportunities for the application of biotechnology. These include development and use of bioherbicides; discovery and use of naturally occurring herbicides; genetic improvement of crop tolerance to herbicides; and use of genetically engineered micro-organisms for other areas of weed management such as decontamination of herbicide spills or protection of selected crops against herbicide injury (Hatzios, 1987; Mukhopadhyay, 1992; Bhowmick and Mandal, 2001).

4.7.1. Development and Use of Bioherbicides:

4.7.1.1. Concept and Basis of the Bioherbicide Strategy:

The term "bioherbicide" refers to microbial plant pathogens which are applied as sprays that uniformly kill or suppress the growth of weeds (Templeton and Smith, 1977). Charudattan and Dinoor (2000) defined "bioherbicide" as a plant pathogen used as a weed-control agent through inundative and repeated applications of its inoculum. In this strategy, microbial plant pathogens are applied to target weeds in a manner similar to chemical herbicide applications (Templeton *et al.*, 1986). Thus, a bioherbicide is a preparation of living inoculum of a plant pathogen, formulated and applied in a manner analogous to that of a chemical herbicide in an effort to control or suppress the growth of weed species (Watson, 1989). The pathogen is not expected to survive or to provide control beyond the growing season in which it is applied (TeBeest and Templeton, 1985).

The use of bioherbicides is based on the fundamental epidemiological principles of plant pathology. Plant disease is the result of the interaction among the host plant, the pathogen and the environment, commonly referred to as the disease triangle. Many factors that often limit disease development include pathogen factors (low inoculum levels, weakly virulent pathogens and poor spore dispersal mechanisms), environmental factors (unfavourable moisture and/or temperature conditions) and plant factors (low susceptibility of the host, host defence mechanisms, and widely dispersed host populations). The bioherbicide approach or strategy is an attempt to bypass many of these restraints on disease development by periodically dispersing an abundant supply of virulent (capable of causing injury) inoculum uniformly onto a susceptible weed population. The application is timed to take advantage of favourable environmental conditions and/or the most susceptible stage of the weed growth. Similarly, the bioherbicide may be formulated to avoid unfavourable environmental conditions and to facilitate application. As a consequence, the development of an effective bioherbicide requires a comprehensive understanding of the pathogen(s) involved, the biology and population dynamics of the target weed(s), the optimum requirements for disease initiation and development, and the complex interactions within host-pathogen system (Watson, 1989).

Fungi, bacteria, and even viruses offer great promise for use as bioherbicides in modern weed management. At present, however, mostly fungal pathogens are in use. Therefore, the term "myco-herbicide" has often been used interchangeably with "bioherbicide" (Scheepens and van Zon, 1982; Templeton *et al.*, 1986; Watson, 1989). In addition, the term "bioherbicide" is generally restricted to the use of plant pathogens and does not include attempts to augment populations of beneficial insects, nor does it generally include the use of naturally occurring compounds (phytotoxins) produced by microorganisms (Watson, 1989). Phyto-pathogenic agents developed as bioherbicides may not be very effective in nature but are made effective for biological control by applying an abnormally high inoculum pressure at an appropriate time (Scheepens and van Zon,1982). Thus, in this strategy, the climax of an epidemic is reached artificially early in the season (Hatzios, 1987) and hence, it is best suited for controlling annual weeds in annual crops where rapid control of weeds is generally desired (Yaduraju *et al.*, 2005). The initial lesions caused by the applied inoculum directly cause the death of infected weed seedlings (TeBeest, 1996).

4.7.1.2. Commercial Prospects and Potentials of Bioherbicides:

The concept of using fungi, bacteria, and even viruses as bioherbicides is biologically feasible with several host-pathogen combinations. Till today, however, mostly fungal pathogens have been exploited commercially as mycoherbicides, marketed for practical applications (Hatzios, 1987; Buttar and Aggarwal, 2005).

The potential commercialization of a microbial phytopathogenic agent greatly depends on whether this microbe possesses properties that allow it to be handled like a chemical herbicide or not. Ideally, a commercialized bioherbicide should be long-lived or insensitive to manipulations in the industrialization process (Kenney, 1986). Thus, determinations of the life cycles of endemic pathogens of major weeds and development of methods for mass production of stable reproductive units (spores) of these pathogens are of prime importance for the commercialization of microbial pathogens as bioherbicides. In addition, the small size of current markets of bioherbicides, their marketability in situations where a selective host range is required, and stabilization of the product under field conditions are factors or obstacles that need to be addressed during the commercialization process (Khachatourians, 1986; Templeton *et al.*, 1986).

A large number of potential biocontrol agents have been identified. However, only a few of them have reached the market place as commercial products or are near to commercialization (Table 3).

DeVine is the first selective and extremely effective mycoherbicide, composed of a Florida isolate of the endemic soil-borne fungus *Phytophthora palmivora* (Butler) Butler, registered in 1981 for the control of strangler (milkweed) vine [*Morrenia odorata* (H.&A.) Lindl] in Florida citrus groves (Kenney,1986; Ridings,1986). *P. palmivora* is a facultative parasite that produces a lethal root rot of its host plant (strangler vine) and persists saprophytically in the soil for extended periods of time (Ridings,1986). Host-specific strains of this fungus are formulated as liquid suspensions consisting largely of 6.7×10^5 live chlamydospores ml^{-1} which are applied at 1.17 l in 450 l of water as sprays with boom and nozzle sprayer to the soil surface under tree canopies (Kenney, 1986; Bhan and Sushilkumar, 1998). In small-scale field tests, 96% of the vines were killed within 10 weeks after infestation of soil with the fungus (Burnett *et al.*, 1973; Ridings *et al.*, 1976 and 1977).

Collego is the second commercially developed mycoherbicide, based on the foliar pathogen *Colletotrichum gloesporoides* (Penz.) Sacc. f. sp. *aeschynomene*, an endemic anthracnose fungus, registered in 1982 for the selective control of a leguminous weed northern jointvetch [*Aeschynomene virginica* (L) B.S.P.] in rice and soybean crops in Arkansas, Louisiana, and Mississippi (Bowers, 1986; Smith, 1986; Templeton *et al*., 1986). Collego is a dry wettable powder containing 15% spores (conidia) of *C. gloesporoides* as active ingredient and 85% inert ingredients. The formulation is rehydrated and resuspended in a sugar medium before mixing with water in the spray tank of an application system (TeBeest

Table 3. List of bioherbicides released commercially or near to commercialization[c]

Product	Microbial agent	Target weed	Crops	Country
DeVine	*Phytophthora palmivora*	*Morrenia odorata* (Strangler vine / milkweed vine	Citrus	U.S.A.
Collego	*Colletotrichum gloesporoides* f.sp. *aeschynomene*	*Aeschynomene virginica* (Northern jointvetch)	Rice & Soybean	U.S.A.
BioMal	*Colletorichum gloesporoides* f. sp. *malvae*	*Malva pusilla* (Round-leaved mallow)	Wheat, flax lentil, mustard & strawberries	Canada
Dr. BioSedge	*Puccinia canaliculata*	*Cyperus esculentus* (Yellow nutsedge)	Horticultural crops	Georgia
Camperico	*Xanthomonas campestris* pv. *poae*	*Poa annua* (Winter bluegrass)	Turf & golf greens	Japan
Stumpout	*Cylindrobasidium laeve*	*Acacia mearnsii & A. pycnantha*	Tree plantations & natural areas	South Africa
Velgo	*Colletotrichum coccodes*	*Abutilon theophrasti* (Velvetleaf)	Corn & Soybean	-
Casst	*Alternaria cassiae*	*Cassia obtusifolia* (Sicklepod)	Soybean, cotton & peanut.	U.S.A.
ABG 5003	*Cercospora rodmanii*	*Eichhornia crassipes* (Water hyacinth)	Waterways	U.S.A.
Luboa-II	*Colletorichum gloesporoides* f.sp *cuscutae*	*Cuscuta* spp. (Dodder)	Soybean, flax, lentil & strawberries	China

[c]After Doraiswamy and Manickam (1996); Yaduraju (2002); Yaduraju *et al.* (2005); Ghorbani *et al.* (2005); Gautam and Mishra (1995); Buttar and Aggarwal (2005).

and Templeton, 1985). In farmers' fields, the commercial formulation of the fungus could provide greater than 90% control of northern joint vetch when used according to label directions (Bowers, 1986). Collego can also be integrated with several chemical herbicides used in soybean, such as acifluorfen and bentazon, in tank-mixed application (TeBeest and Templeton, 1985). However, this mycoherbicide has not been marketed by the registrant since 1992 (TeBeest, 1996)

The third mycoherbicide, BioMal, containing spores of *C. gloesporoides* (Penz.) Sacc. f. sp. *malvae* was registered in 1992 in Canada for the control of round-leaved mallow (*Malva pusilla* Sm.) in wheat. Round-leaved mallow plants were killed within 17 to 20 days after inoculation with spore suspensions. Control was 90 to 100% effective in the field and the infected plants did not reappear in the following year. Though registered, BioMal has not been available commercially (TeBeest, 1996). But it is currently under development with a different formulation from BioMal and under a new commercial name of Mallet WP (Ghorbani *et al.*, 2005).

Dr. BioSedge is a mycoherbicide, based on the rust fungus *Puccinia canaliculata* (Sacho.) Lagerh. and registered for the control of yellow nutsedge (*Cyperus esculentus* L.), presently unavailable for commercial use (Ghorbani *et al.*, 2005). It is used to control nutsedge weeds in horticultural crops and application rate is 5 mg uredospores ha^{-1}. This fungus is capable of rapid dispersal and infection, and can cause as high as 90% reduction in weed biomass. Early application also inhibits new tuber formation up to 66% (Yaduraju *et al.*, 2005).

Camperico is a bioherbicide, containing an isolate of wilt-inducing bacterium *Xanthomonas campestris* pv. *poae,* to control winter bluegrass (*Poa annua* L.) in golf courses (Ghorbani *et al.*, 2005; Yaduraju *et al.*, 2005; TeBeest, 1996). Johnson (1994) reported that three applications of two strains of this bacterium controlled 52 to 82% of the bluegrass in dormant bermudagrass [*Cynodon dactylon* (L.) Pers.] field plots.

Stumpout is another bioherbicide, based on the wood-infecting basidiomycete, *Cylindrobasidium laeve*, registered in South Africa to control resprouting of cut trees in tree plantations and in natural areas (Charudattan, 2001; Charudattan and Dinoor, 2000; Ghorbani *et al.*, 2005).

Gupta *et al.* (2002) demonstrated the potential of *Puccinia romagnoliana* Marie & Sacc. as a bioherbicide to control the purple nutsedge (*Cyperus rotundus* L.) in mulberry fields. Karuppiah and Seetharaman (2004) first in India recorded the potential of rust fungi *Puccinia* spp. as bioherbicide in controlling purple nutsedge (*Cyperus rotundus* L.), umbrella sedge (*C. difformis* L.), korai (*C. flavidus* Retz.) and hoorah grass [*Fimbristylis miliacea* (L.) Vahl.], and smut fungal pathogen *Cintractia peribebuyensis* to control rice flat sedge (*Cyperus iria* L.) and purple nutsedge (*C. rotundus* L.) in rice.

There are several prospective mycoherbicides, of which Velgo [*Colletotrichum coccodes* (Wallr.) Hughes] to control velvetleaf (*Abutilon theophrasti* Medic.), Casst (*Alternaria cassiae* Jurair & Khan) to control sicklepod weed (*Cassia obtusifolia* L.) in cotton & soybean, ABG 5003 (*Cercospora rodmanii* Conway) to control water hyacinth [*Eichhornia crassipes* (Mart.) Solms] in waterways, and Luboa-II [*Colletotrichum gloesporoides* (*Glomerella cingulata*) f.sp. *cuscutae*] to control dodder (*Cuscuta* spp.) have already shown great promise (Mukhopadhyay, 1993; Yaduraju *et al.,* 2005). These milestones have resulted in the heightened search for more and more organisms with potentials for controlling weeds throughout the world. Thus, a number of other pathogens are in various stages of research and development as bioherbicides. Some of these are listed in Table 4.

Table 4. List of pathogens under consideration or investigation as bioherbicides[d]

Plant Pathogen	Target weed	Crop	Country
Alternaria crassa	Jimson weed (*Datura stramonium)*	Soybean	U.S.A.
Alternaria macrospora	Spurred anoda (*Anoda cristata)*	Cotton	U.S.A.
Ascochyta caulina	Bathua (*Chenopodium album)*	Wheat, sugarbeet, maize & pea	Europe
Bipolaris sorghicola	Johnson grass (*Sorghum halepense)*	-	U.S.A.
Chondrostereum purpureum	Black cherry (*Prunus serotina*)	Forest	Netherlands
Colletotrichum dematium	Tall morning glory (*Ipomoea purpurea)*	-	U.S.A
Colletotrichum gloesporoides	*Hakea serecea*	-	South Africa
Colletotrichum malvarum	Prickly sida (*Sida spinosa*)	Cotton & soybean	U.S.A
Colletotrichum orbiculare	Bathurst bur (*Xanthium spinosum*)	-	Australia
Colletotrichum truncatum	Hemp sesbania(*Sesbania exaltata*)	Soybean & rice	U.S.A.
Dendryphiella sp.	Water chestnut (*Eleocharis kuroguwai)*	Rice	Japan
Drechslera monoceras	Barnyard grass (*Echinochloa crusgalli*)	Rice	Japan
Epicoccosorus nematosporus	Water chestnut (*Eleocharis kuroguwai*)	Rice	Japan
Exserohilum monoceras	Barnyard grass (*Echinochloa* spp.)	Rice	-
Fusarium moniliforme	Jimson weed (*Datura stramonium*)	Rice	Mississippi
Fusarium oxysporum f.sp. *orthoceras*	*Orobanche cumana*	Sunflower	Bulgaria

Table 4. (Continued)

Fusarium tumidum	Gorse (*Ulex europaeus) &* broom *(Cytirus scoparius)*	Maize & pearl millet	-
Phomopsis amaranthicola	*Amaranthus* spp.	-	U.S.A.
Phragmidium violaceum	Black berry (*Rubus* spp.)	-	Australia
Pseudomonas syringae pv. *tagetis*	Canada thistle (*Cirsium arvense*)	Soybean	U.S.A.
Puccinia carduorum	Musk thistle (*Carduus nutans*)	-	U.S.A.
Puccinia chondrillina	Rush skeletonweed (*Chondrilla juncea*)	Wheat	Australia
Sclerotinia sclerotiorum	Canada thitle (*Cirsium arvense*)	Pastures	U.S.A.
Sclerotinia sclerotiorum	*Ranunculus acris*	Dairy pasture	-
Sphacelotheca holci	Johnson grass (*Sorghum halepense*)	-	U.S.A.

dAfter Doraiswamy and Manickam (1996); TeBeest (1996); Yaduraju et al. (2005); Ghorbani et al. (2005); Buttar and Aggarwal (2005)

Recent efforts on the bioherbicide development are being concentrated on the use of obligate fungal pathogens, non-phytopathogenic fungi and bacteria, phytopathogenic bacteria, and nematodes (Kremer, 2002; Rosskoph *et al.*, 1999) as listed in Table 5.

Despite intense research efforts on bioherbicides, only eight or nine commercial products are practically in use worldwide. Among all the potential microbial agents evaluated for weed management, though the rate of success is very small, being only around 10%, these have had a high rate of return on scientific and monetary inputs as compared to chemical herbicides (Yaduraju *et al.*, 2005).

Efforts are also being made to develop more efficient strains or isolates of bioherbicide pathogens and other genetic manipulations like recombinant DNA technology to enhance bioherbicide performance. Enhancement of bioherbicide potential may also be obtained with the combination of growth regulators and chemical herbicides. For example, tank mix applications of *C. coccodes* and a growth regulator, thidiazuron acted synergistically to increase velvetleaf mortality when compared with the fungus alone (Wymore *et al.*, 1987; Hodgson *et al.*, 1988). Combined use of the herbicide pyrazosulfuron ethyl and the fungus *Drechslera monoceras* showed significant synergism in controlling barnyard grass in Japan (Gohbara and Yamaguchi, 1993). Thus, bioherbicides are not seen as alternatives to chemical herbicides but as complementary adjuncts to current weed management systems (Bhowmick and Biswas, 2007).

Table 5. Some promising groups of microbes as potential bioherbicides[e]

Bioagent	Target weed	Targeted ecosystem	Current status
1. Soil-borne fungi Sclerotinia minor	*Taraxacum officinale* (Dandelion)	Turf	Under field testing & formulati development
Fusarium oxysporum f.sp. *orthoceras*	*Orobanche cumana*	Sunflower	Small-scale field trials
2.Non-phytopathogenic soil-borne fungi *Gliocladium virens*	Several	Row crops	Under testing for optimal toxin production, delivery & field testing
Trichoderma spp.	*Phalaris minor*	Wheat	Under development
3.Phytopathogenic bacteria *Pseudomonas syringae* pv. *phaseolicola*	*Pueraria lobata* (Kudzu)	Pastures	Field testing
4.Non-phytopathogenic bacteria (Deleterious rhizobacteria) *Pseudomonas fluorescens* D7	*Bromus tectorum* (Downy brome)	Cereal Crop	Under Development
Enterobacter	*Bromus tectorum*	Cereal Crop	Large scale field testing
Pseudomonas	*Bromus japonicus*	-	-
Xanthomonas	*Aegilops cylindrica*	-	-
Agrobacterium	*Euphorbia esula*	Rangeland	Small scale field testing
5.Nematodes *Subanguina picridis*	*Acroptilon repens* (Russian knapweed)	Rangeland	Under field testing & formulation evaluation

[e]After Yaduraju *et al.* (2005).

Templeton *et al.* (1986) suggested that bioherbicides would have the greatest potential in the following areas that are currently served inadequately by chemical herbicides:

1. Control of parasitic weeds: Parasitic weeds (phanerogamic plant pathogens) have been a major problem in many crop production regions throughout the world, causing drastic yield reductions. There are no selective herbicides for satisfactory control of these weeds. So, bioherbicides could be potentially effective on these weeds. The success of *Colletotrichum gleosporoides* f.sp. *cuscutae* on *Cuscuta* spp. in China and *Fusarium oxysporum* f.sp. *orthoceras* on *Orobanche cumana* in Bulgaria are milestones in this direction (Yaduraju *et al.*, 2005).

2. Control of weeds closely related to crops (crop mimics), requiring a high degree of selectivity: In some cases, crop and weed seedlings look similar and cannot be differentiated easily even by skilled labourers. For example, *Phalaris minor* in wheat and *Echinochloa crusgalli* in rice (Mukhopadhyay, 1992 and 1997). Selective bioherbicides may be a good answer in taking care of such problems.
3. Control of weeds resistant to chemical herbicides: Evolution of herbicide-resistant weed biotypes either due to repeated use of single herbicide in conventional system or due to repeated use of a broad-spectrum non-selective herbicide in a herbicide-resistant transgenic crop has become a challenge to modern agriculture. The bioherbicide strategy may offer an adequate control of these weeds. In India, isoproturon resistant biotypes of *Phalaris minor* might be a good target for strains of many *Trichoderma* spp. which are under investigation (Yaduraju *et al.*, 2005).
4. Control of weeds infecting small specialized areas where development of chemical herbicides would be too costly.

4.7.1.3. Factors to be considered in Selecting Biocontrol Agents as Bioherbicides:

In the development of any new pest control strategy, safety and efficacy are the two primary concerns. As a consequence, safety (in relation to crop plants, environment, and human health) and efficacy (in relation to environmental tolerance, level of damage to the weed, and ability to be integrated within the crop production system) are the major criteria in the selection of suitable plant pathogens (Watson, 1989).

Fungal pathogens that are virulent, host-specific, and genetically stable but constrained naturally by low inoculum production and poor dissemination are good candidates for development as mycoherbicides (Templeton *et al.*, 1986; Buttar and Aggarwal, 2005). The exploitation of plant pathogenic bacteria as bioherbicides seems feasible because they can be host-specific and can easily be cultivated *in vitro* (Scheepens and van Zon, 1982; Buttar and Aggarwal, 2005). In contrast, plant pathogenic viruses are more problematic since they are often not host-specific and a vector is usually necessary for their transmission from one plant to another. Other factors influencing the potential use of plant pathogens as bioherbicides include low virulence, stringent temperature and moisture requirements, wounding requirements, and specific physiological interactions with the host plant (TeBeest and Templeton, 1985).

However, the preferred characteristics of a potential bioherbicide pathogen include: (1) growth and sporulation on artificial media, (2) highly virulent against the target, (3) genetic stability, (4) restricted host range, (5) broad tolerance range, (6) prolific propagule production, (7) capacity to damage its host plant, and (8) innocuous in ecological effects (Templeton *et al.*, 1979). In addition, they must be safe, as they are released in the field at high population densities (Charudattan *et al.*, 1985). Shelf-life is also very important for a commercial bioherbicide; the longer the shelf-life of the product, the greater the chance that it will be developed, registered, and commercialized (Ghorbani *et al*, 2005).

4.7.1.4. Bioherbicide Formulations:

Biocontrol agents should be formulated appropriately for increasing their efficacy against environmental constraints. However, these are not easy to formulate into effective products. As these are living organisms, their viability must be preserved throughout processing and storage (Auld *et al.*, 2003; Hoagland, 1990; Scher and Castagno, 1986). Perhaps the biggest single constraint for the commercial development and marketing of bioherbicides is development of an appropriate formulation (Greaves *et al.*, 1999).Biocontrol agents may be formulated either as liquid sprays or as solid materials (Ghorbani *et al.*, 2005).

Liquid formulations of mycoherbicides are usually the best suited for use as post-emergence sprays and are used primarily to incite leaf and stem diseases. Compared to fungal bioherbicides (mycoherbicides), bacterial pathogens such as *Xanthomonas campestris* pv. *poae* and *Pseudomonas syringae* pv. *tagetis* are very easy and quick to produce in liquid cultures and to store as frozen or freeze-dried pellets. Even simpler to produce, store, and use are mechanically transmitted viruses such as *Araujia* mosaic virus and tobacco mild green mosaic virus U2, which can be multiplied on tolerant (nonlethal) hosts, freeze dried, and stored for several years without loss of infectivity (Charudattan, 2001). Conversely, pathogens that infect at or below the soil surface are probably best delivered in a solid or granular formulation. Granular formulations are often better suited for use as pre-plant or post-emergence mycoherbicides than spray formulations because the granules provide a buffer from environmental extremes; the granules can serve as a food base for the fungus and are less likely to be washed away from treated areas than spores in a liquid media (Boyette and Abbas, 1995).

4.7.1.5. Advantages of Bioherbicides:

The utilization of bioherbicides in weed management would have the advantages (Khachatourians, 1986; Templeton *et al.*, 1986; Gautam and Mishra, 1995; Yaduraju and Mishra, 2005; Mukhopadhyay, 1997; Buttar and Aggarwal, 2005), given here under:

1. A high degree of specificity for the target weed;
2. No effects on non-target and beneficial plants or man;
3. Absence of weed resistance development;
4. Absence of residue build up in the environment; and
5. Potential impact from biotechnological research and development.

Moreover, the cost-benefit analysis is more favourable for bioherbicides than for chemical herbicides, as capital outlays needed for research, development, and registration of the latter is more than that in case of the former (Charudattan, 2001; Ghorbani *et al.*, 2005). For example, the cost of developing Collego has been estimated to be \$1.0 to 1.5 million, being substantially less than that for a chemical pesticide which normally costs between \$25 to 50 million (Templeton *et al.*, 1986). The development of new herbicide chemicals has become more expensive with additional chemical regulations, more stringent toxicology standards and increased concern for the environment (Chu, 1983).

4.7.1.6. Disadvantages of Bioherbicides:

Despite having several advantages, bioherbicides have some disadvantages, constraints or limitations. As with other biocontrol agents, these are difficult to formulate and this is the primary technological constraint on the development of bioherbicides (Ghorbani *et al.*,2005). Besides, there are some drawbacks associated with the production and usage of bioherbicides (Khachatourians, 1986; Templeton *et al.*, 1986; Buttar and Aggrawal, 2005). These are as follows:

1. Bioherbicides have to be registered with the Environmental Protection Agency (E.P.A.) and the registration process may be lengthy. Even the market for bioherbicides is not big enough to earn back the registration costs in a reasonable period of time (Scheepens *et al.*, 2001).
2. Suppression or killing of weeds by bioherbicides may be a slow process.
3. Stability of bioherbicides under field conditions is highly dependent on environmental conditions.
4. Production of a bioherbicide for large-scale application may be an expensive process.
5. Numerous fungi need to be discovered and developed as bioherbicides because of their high degree of specificity.

However, the major difficulties with the bioherbicide strategy are the needs for virulent and host-specific pathogen/bioagent, dependence upon environmental conditions for maximum damage (development of epidemic), requirement of ideal conditions for storage and transport of bioagent, and limited shelf-life of bioagent (Yaduraju *et al.*, 2005).As the "active ingredient" of bioherbicides is a living organism, extra care is required in shipping, storage, and handling (Ghorbani *et al.*, 2005).In addition, numerous precautions that must be taken while using bioherbicides in the field include the prevention of exposure of these formulations to wetting agents, fertilizers and chemical pesticides that are detrimental to the viability of their spores

4.7.2 Discovery and Use of Naturally Occurring Herbicides:

As a consequence of worldwide growing concern about the environmental aspects of selected pesticides, the need for the use of easily degradable pesticides with good selectivity is greater than ever. Advances in microbial and plant biochemistry have stimulated scientific interest into the possible role of secondary plant products and microbial toxins as natural pesticides. Furthermore, advances in plant cell culture, fermentation technology, molecular genetics and genetic engineering make it possible to exploit biotechnologically plants and micro-organisms as potential sources of naturally occurring chemical compounds that could be developed as herbicides (Hatzios, 1987). Extensive research during the last few decades has demonstrated that several plant secondary metabolites (allelochemicals) as well as fungal and microbial toxins possess good herbicidal activity (Bhowmick and Mandal, 2001).

Of the naturally occurring compounds, bialophos is marketed under the trade name Herbiaceae as a commercially developed herbicide in Japan (Mase, 1984; Tachibana, 1987). Phosphinothricin is the active ingredient of this microbial herbicide bialophos. It is a microbial product isolated from the fermentation broth of *Streptomyces hygroscopicus* and *S. viridochromogenes*, and exhibits strong herbicidal activity against a wide spectrum of grass and broadleaf weeds following application to their foliage (Mase, 1984; Misato and Yamaguchi, 1984). Since bialophos has the remarkable biodegradability by soil micro-organisms, it shows very little or no toxicity when applied to soil (Misato and Yamaguchi, 1984). Phosalacine is another microbial compound containing phosphinothricin and its herbicidal behaviour is similar to that of bialophos (Omura *et al.*, 1984). A number of other natural products from higher plants and micro-organisms including fungi (tentoxin, picloram, anisomycin, etc.) are in various stages of development for agricultural exploitation in various ways.

The problems of high phytotoxicity, limited crop selectivity, and instability under field conditions associated with several naturally occurring compounds hinder their commercial development as herbicides. These problems can be overcome by a biorational synthesis of more selective and stable analogs of these chemicals. Thus, microbial toxins and allelochemicals provide us with novel chemistries that could be manipulated in order to produce commercial herbicides (Hatzios, 1987). Selected examples of commercially developed herbicides which are chemically similar to naturally occurring phytotoxic compounds are listed in Table 6.

Table 6. Commercially developed herbicides based on natural chemistry[f]

Natural product	Plant or microbial source	Herbicide	Manufacturer /country
Anisomycin	*Streptomyces* sp.	Methoxyphenone	Nihon/Japan
Cineole	Widespread in plants	Cinmethylin	Shell/U.S.A.
Benzoxazinones (hydroxamic acids)	Poaceous plants	Benzazin	BASF/ Federal Republic of Germany
Iprexil	*Iprex pachyon*	Benzadox	Gulf/U.S.A.
Fusaric acid	*Fusarium* sp.	Picloram	Dow/U.S.A.
Moniliformin	*Fusarium moniliforme*	3,4-dibutoxy-moniliformin	CIBA-GEIGY/ Switzerland
Quinolinic acid	*Nicotiana tabacum*	Quinclorac	BASF/ Federal Republic of Germany
Phosphinothricin	*Streptomyces viridochromogenes*	Glufosinate	Hoechst/ Federal Republic of Germany

[f]Hatzios (1987)

The development of the herbicide NK-049 (Methoxyphenone) as a synthetic analog of the microbial toxin anisomycin from *Streptomyces* sp. represents one of the few successful efforts employing the biorational chemical synthesis of a herbicide based on a natural chemistry (Munakata *et al.*, 1973). Methoxyphenone is marketed in Japan as a selective herbicide for the control of barnyard grass [*Echinochloa crusgalli* (L.) Beauv.] in rice and is easily degraded in soil (Duke, 1986; Munakata *et al.*, 1973). The ammonium salt of glufosinate, a synthetic racemic mixture based on the chemistry of phosphinothricin, has been introduced as a non-selective herbicide marketed under the trade name Basta (Kocher, 1983). These examples demonstrate that the structures of naturally occurring phytotoxins can serve as leads for the synthesis of new successful herbicides.

4.7.3. Genetic Improvement of Crop Tolerance to Herbicides:

It is clear that the selective herbicide can manage only a selection of weeds, whereas non-selective herbicide kills all the plants including the crop. This problem of crop susceptibility can be solved by screening tolerant available cultivars and developing resistance in crops against broad spectrum herbicides, which provide an effective weed management. Differential tolerance of herbicides has been noted in several crops as in weeds, indicating a great possibility for success in breeding hervicide tolerant cultivars.

To generate herbicide resistant crop plants, four major approaches possible (Hatzios, 1987) are classic plant breeding with herbicide resistant biotypes and mutation breeding including mutagenesis of seeds followed by screening of seedlings for resistance; *in vitro* mutant selection (by culturing cells or tissues in normally toxic concentrations of herbicides); mutant selection by somatic hybridization (fusion of protoplasts in culture from different plants to combine genetic information to create a new hybrid); and transfer of cloned genes into susceptible crop plants (through the use of vectors or protoplasts).

In the United States, IMI-Corn (tolerant to imidazolinone herbicides) was introduced in 1991 as the first major transgenic crop. Other herbicide tolerant crops are SR-Corn (tolerant to sethoxydim), Liberty Link Corn (resistant to glufosinate), STS Tolerant Soybeans (sulfonylurea tolerant), Roundup Ready Corn (tolerant to glyphosate) and Roundup Ready Soybeans (tolerant to glyphosate) as reported by Bhowmik (1999).

Herbicide tolerant cultivars have three obvious advantages. Firstly, it is possible to use higher rates of herbicides to eliminate hard-to-control weeds. Secondly, it provides for more flexible use of herbicides, and results in saving of time and labour. Thirdly, there is always protection of valuable crops even when accidentally treated with non-selective herbicides (Gautam and Mishra, 1995).

4.7.4. Additional Uses of Genetically Engineered Micro-organisms in Weed Management:

Besides various uses of microorganisms in selected areas of weed management (e.g., bio-herbicides, production of natural phytotoxins, and genetic transformation of plants), genetically engineered microorganisms may be useful for solving a number of other difficulties ivolved in modern weed management and herbicide technology. These areas include microbial activity with the help of plasmid borne gene systems - biodegradation or mineralization of herbicide in water and soil; development and use of microbial herbicide safeners; and use of microorganisms as biocatalysts in the production of synthetic herbicides (Hatzios, 1987).

4.7.4.1. Plasmid-borne Gene Systems that Degrade Soil Applied Herbicides:

Biodegradation is a very important factor affecting the environmental persistence of most organic herbicides used in modern agriculture. Biodegradation or mineralization of herbicides in water and soils is almost always the result of microbial activity. Soil microbes utilizer either adaptive or constitutive enzyme systems which can alter biochemically a pesticide into a utilizable nutrient and energy source. A number of herbicides are completely mineralized by selected soil micro-organisms and specific examples are presented in Table 7.

Table 7. Microbial strains known to metabolize selected herbicides in soils[g]

Herbicide	Microbe
Phenoxyacetates (2,4-D, MCPA, 2,4,5-T)	*Alcaligenes* sp., *Achromobacter* sp., *Arthrobacter* sp., *Aspergillus* sp., *Brevibacterium* sp., *Corynebacterium* sp., *Nocardia* sp., *Pseudomonas* sp., *Streptomyces* sp.
Chlorpropham	*Aspergillus* sp., *Pseudomonas* sp.
Pentachlorophenol	*Arthrobacter* sp., *Flavobacterium* sp., *Pseudomonas* sp.
Pyrazon	*Phenylobacterium* sp.
Dalapon, Glyphosate	*Pseudomonas* sp.
Alachlor	*Chaetomium globosum, Rhizoctonia solani, Fusarium oxysporum*
Metolachlor	Unspecified actinomycete, *Chaetomium globosum, Bacillus circulans, B. megatherium, Fusarium* sp., *Mucor racemosus*
Propanil	*Fusarium solani*
Phenylureas	*Bacillus* sp., *Pseudomonas* sp.
EPTC	*Arthrobacter* sp., Fluorescent *Pseudomonads*
Molinate	*Mycobacterium* sp., *Flavobacterium* sp., *Streptomyces* sp., *Fusarium* sp.

[g]Hatzios (1987)

4.7.4.2. Microbial Herbicide Safeners:

Microbes carrying plasmids with herbicide degradative genes could be used as crop protectants to increase the selectivity of soil-applied herbicides. Genes encoding for degradative enzymes which break down specific herbicides, isolated from plants or microbes, could be cloned into appropriate plasmids and transferred to bacteria (e.g., *Pseudomonas fluorescens*) that colonize the roots or seeds of susceptible plants from soil-applied and root-absorbed herbicides (Hatzios, 1987).

An alternative approach for the use of micro-organisms as safeners for the protection of crop plants against herbicide injury could include the direct insertion of microbial genes encoding for herbicide degradative enzymes into the chromosomes of susceptible plants. The goal of this approach is to a producer of the desirable enzyme which provides protection against injury from a given herbicide (Hatzios, 1987).

4.7.4.3. Use of Microorganisms as Biocatalysts in the Production of Synthetic Herbicides:

Apart from their potential use as bioherbicides or as sources of novel phytotoxins and leads for new herbicide chemistries, micro-organisms could be further exploited as biological catalysts for the synthesis and commercial production of sophisticated plant protection chemicals (Hatzios, 1987).

Integrated Approach in Weed Management

Integrated weed management (IWM) is a weed management system that uses all suitable weed management approaches / methods in a compatible manner so as to reduce weed population and maintain them at level below those causing economic injury. Modern trend in the field of weed management is to promote IWM so that chemical for weed management is to be used as minimum as possible and as a last resort. Bhan and Singh (1993), and Singh (2001) reported this aspect at length.

5. Weed Problems – A Crop-Wise Analysis

Weed problems vary from crop to crop, region to region, and farm to farm. Also, the spectrum may vary with soil types. As weeds often have a higher rate of initial growth, they compete very efficiently with the crops in early stages of growth. Therefore, weeds, if not properly managed, remove huge amount of soil moisture and nutrients, and deprive the crop of the benefit of moisture and nutrients inherent in soil as well as added fertilizers and irrigation water meant for crop, finally causing reduction in crop yield (Table 2). The critical period of crop-weed competition has been presented in Table 8. Each crop as well as non-crop area has also a particular status with respect to its associated weed flora (grasses/ sedges/ broad-leaved weeds), as briefed hereunder.

Table 8. Critical period of crop-weed competition in important field crops of India

Crop	**Critical period of competition**	**Crop**	**Critical period of competition**
Direct seeded rice	15-30 DAS	Chickpea	30-60 DAS
Tranplanted rice	30-45 DAT	Peas	30-45 DAS
Wheat	30-50 DAS	Soybean	30-50 DAS
Jute	First 6 weeks	Groundnut	40-60 DAS
Sugarcane	40-70 DAP	Rapeseed-mustard	15-40 DAS
Potato	First 4 weeks	Sesame	15-45 DAS
Maize	30-40 DAS	Linseed	20-45 DAS
Greengram	15-45 DAS	Safflower	15-45 DAS
Blackgram	15-45 DAS	Sunflower	15-40 DAS
Pigeonpea	15-60 DAS	Castor	30-60 DAS
Cowpea	15-35 DAS	Niger	15-45 DAS
Lentil	30-60 DAS	Vegetables	30-35 DAP

DAP: Days after planting; DAS: Days after sowing; DAT: Days after transplanting

5.1. Direct seeded upland rice:

Complete crop failure is not uncommon due to weed infestation in upland direct seeded upland rice (Mukhopadhyay, 1992). Major weeds are *Cynodon dactylon* (Durba), *Echinochloa colona* (Shyama), *Eleusine indica* (Kodai), *Cyperus rotundus* (Motha), *Ageratum conyzoides* (Chhagronya) and *Commelina benghalensis* (Kansira).

5.2. Transplanted lowland rice:

In general, major weeds in transplanted lowland rice (Bhowmick, 2001; Bhowmick *et al.*, 2005) are *Echinochloa crusgalli* (Shyama), *Eleusine indica* (Kodai), *Leersia hexandra* (Kara), *Cyperus rotundus* (Motha), *C. difformis* (Motha), *C. iria* (Jal motha) *Fimbristylis littoralis* (Bindi motha), *Ammania baccifera* (Dadmari), *Ludwigia parviflora* (Labangaphul),

Marsilea quadrifolia (Shusni), *Monochoria vaginalis* (Panari) and *Sphenoclea zeylanica* (Hansi ghas). Ghosh *et al.* (2005) reported that the prominent weeds in transplanted *kharif* rice were *Echinochloa glabrescens, Leersia hexandra, Cyperus difformis, Fimbristylis littoralis, Ammania baccifera, Alternanthera philoxeroides* and *Marsilea quadrifolia.* Algal weeds common under lowland situation are *Chara, Nitella*, etc.

5.3. Wheat:

Major weeds in wheat are *Cyperus rotundus* (Motha), *Anagallis arvensis* (Krishnaneel), *Chenopodium album* (Bathua) and *Melilotus alba* (Safed senji). Weeds are rare in West Bengal. but common in Punjab, Haryana, Uttar Pradesh and Rajasthan are *Avena fatua* (Bunojui), *A. ludoviciana* (Jangli sp.), *Phalaris minor* (Phala ghas) and *Lolium temulentum* (Karat ghas).

5.4. Jute:

In jute, there may be loss of fibre yield to an extent of 80%, if no weeding is done (Saraswat and Mitra, 1977). The common weeds observed in jute field are *Cynodon dactylon* (Durba), *Echinochloa colona* (Shyama), *Eleusine indica* (Kodai), *Cyperus rotundus* (Motha), *Melochia corchorifolia* (Tiki okra), *Euphorbia hirta* (Dudhia), *Phyllanthus niruri* (Hazardana) and *Physalis minima* (Bantepari).

5.5. Sugarcane:

The sugarcane crop gets infested with a variety of weeds (Bhowmick, 2004), of which the major ones are *Cynodon dactylon* (Durba), *Imperata cylindrica* (Ulu), *Sorghum halepense* (Banchari), *Cyperus rotundus* (Motha), *Euphorbia hirta* (Dudhia), *Physalis minima* (Bantepari) and *Solanum nigrum* (Gurki). If uncontrolled, they rob off about 40% of the nutrients applied to the crop. Crop losses due to weeds vary from 10 to 70% in terms of yield reduction, depending upon the type and intensity of weed flora present in an area.

5.6. Potato:

In the Gangetic alluvial zone of West Bengal (India), weeds cause about 78% yield reduction in potato (Ghosh *et al.*, 2002). The predominant weeds (Ghosh *et al.*, 1999 and 2002) are *Cyperus rotundus* (Motha), *Chenopodium album* (Bathua), *Digitaria sanguinalis* (Keoai), *Eleusine indica* (Kodai), *Melilotus alba* (Safed senji), *Physalis minima* (Bantepari) and *Portulaca oleracea* (Numiasak) and *Eclipta alba* (Keshuti).

5.7. Maize:

Major weeds comprise of *Cynodon dactylon* (Durba), *Echinochloa colona* (Shyama), *Eleusine indica* (Kodai), *Cyperus rotundus* (Motha), *Euphorbia hirta* (Dudhia), *Eclipta alba* (Keshuti) and *Portulaca oleracea* (Numiasak).

5.8. Pulses:

Pulse crops get infested with numerous weed species, of which the major ones are *Cynodon dactylon* (Durba), *Digitaria sanguinalis* (Keoai), *Cyperus rotundus* (Motha), *Chenopodiun album* (Bathua), *Cleome viscose* (Jangli hurhur), *Ageratum conyzoides* (Chhagronya), *Melilotus alba* (safed senji), *Vicia sativa* (Hetka) and *Lathyrus aphaca* (Jangli motor). The yield losses in pulses due to weeds vary from 26.9% to as high as 82.2% (Saraswat and Mishra, 1996).

5.9. Oilseeds:

One of the major constraints to higher production in oilseeds is the heavy infestation of weeds and their ineffective management (Upadhyay, 1985; Saraswat and Mishra, 1996). Major weeds are *Cynodon dactylon* (Durba), *Echinochloa colona* (Shyama), *Pennisetum pedicellatum*, *Cyperus rotundus* (Motha), *Cleome viscose* (Jangli hurhur), *Euphorbia hirta* (Dudhia) and *Physalis minima* (Bantepari).

5.10. Vegetables:

Major weeds infesting vegetable crops include *Echinochloa colona* (Shyama), *Cyperus rotundus* (Motha), *Melilotus alba* (Safed senji), *Amaranthus spinosus* (Kantanote) and *Anagallis arvensis* (Krishnaneel).

5.11. Non-crop areas:

In recent years, the increasing trends of some pernicious weeds in non-crop areas jeopardize the natural environment. These weeds are most neglected although they develop unsightly and dangerous growth (Bhowmick *et al.*, 2005). There is no better example to this in India than the way *Parthenium hysterophorus* L. is flourishing and becoming a serious menace in vacant and uncultivated areas including roadsides, rail roads, industrial sites and airfields (Bhowmick, 2000; Bhowmick *et al.*, 2005). Major weeds in these areas (Ghosh *et al.*, 2002; Bhowmick *et al.*, 2005) mainly comprise of *Cynodon dactylon* (Durba), *Digitaria sanguinalis* (Keoai), *Eleusine indica* (Kodai), *Imperata cylindrica* (Ulu), *Paspalum conjugatum* (Paspalum), *Cyperus rotundus* (Motha), *Ageratum conyzoides* (Chhagronya), *Parthenium hysterophorus* (Parthenium) and *Heliotropium indicum* (Hatisur).

6. RECOMMENDATIONS ON WEED MANAGEMENT

6.1. Direct seeded upland rice:

Different options for weed management are as follows:

a. Cultural: Deep summer ploughing, timely sowing, use of higher seed rate (80-100 kg/ha), line sowing with 20 cm row spacing and stale seed bed technique.
b. Mechanical: Hand weeding twice at 20-25 and 40-45 days after sowing
c. Chemical: Herbicides found promising are presented in Table 9.

Table 9. Promising herbicides for weed management in direct seeded upland rice

Herbicide	Dose (Kg a.i./ha)	Time of application	Weed spectrum
Butachlor 50 EC	1.5	PE (4-6 DAS)	G
Oxadiazon 25 EC	1.0	PE	G+BL
Pendimethalin 30 EC	1.0-1.5	PE (0-6 DAS)	G+BL
Oxyfluorfen 23.5 EC	0.15	PE	G+S+BL
Ethoxysulfuron 15 WSG	0.02	POE (10-12 DAS)	S+BL
Metsulfuron methyl + Chlorimuron ethyl 20 WP	0.004	POE (15 DAS)	S+BL
2,4-D Na Salt 80 WP	1.0	POE	S+BL

G: Grass; S: Sedge; BL: Broad leaf; DAS: Days after sowing; PE: Pre-emergence; POE: Post-emergence

6.2. Transplanted lowland rice:

Different options for weed management are as follows:

a) Cultural: 1. Transplanting of rice seedlings free from weed seedlings
2. Maintenance of submerged condition.
3. Keeping bunds & irrigation channels clean.

b) Mechanical: Hand weeding (HW) twice at 20 and 40 days after transplanting (DAT).

c) Chemical: Ethoxysulfuron 15 WSG when applied at 0.02 kg/ha as post-emergence (POE) effectively kept most of the sedges and broad leaf weeds under control and improved both the grain and straw yields. The same chemical at the same dosage when tank-mixed with anilofos 30 EC at 0.375 kg/ha and applied as POE had a wide efficacy spectrum with regard to the management of all the categories of dominant weeds as well as complex of weed flora and found as effective as HW twice at 20 and 40 DAT during both the *boro* (Bhowmick *et al.*, 2000; Bhowmick, 2001a; Bhowmick and Ghosh, 2002b and 2006) and *kharif* (Pal *et al.*, 2002) seasons. Pre-emergence application of pretilachlor at 500 g/ha alone or at 437.50 g/ha in combination with one HW could help the crop plants to better utilize the solar energy by curtailing all sorts of weed growth and gave results comparable to two rounds of HW (20 and 40 DAT) in transplanted *kharif* rice (Islam *et al.*, 2001; Bhowmick, 2002a; Bhowmick *et al.*, 2005). Application of clomazone in combination with propanil at lower dose as early POE (10 DAT) or at higher dose as late POE (20 DAT) exhibited better weed management, recorded higher crop yields and gave results comparable to HW twice in transplanted puddled rice during *kharif* (Ghosh *et al.*, 2002). However, the herbicides found promising in transplanted rice are given in Table 10.

Table 10. Promising herbicides for weed management in transplanted lowland rice

Herbicide	Dose (Kg a.i./ha)	Time of application	Weed spectrum
Butachlor 50 EC	1.500	PE (3-5 DAT)	G
Oxadiargyl 80 WP	0.100	PE (3-5 DAT)	G+BL
Pendimethalin 30 EC	1.000	PE (1-3 DAT)	G+BL
Oxyfluorfen 23.5 EC	0.150	PE (4 DAT)	G+S+BL
Pyrazosulfuron ethyl 10 WP	0.150-0.200 0.300	PE (3-4 DAT) POE (15 DAT)	S+BL
Pretilachlor 50 EC	0.500	PE (3-5 DAT)	G+S+BL
Pretilachlor 50 EC + One HW	0.400 -	PE (3-5 DAT) 40 DAT	G+S+BL
Butachlor 50 EC fb. 2,4-D Na salt 80 WP or Metsulfuron methyl + Chlorimuron ethyl 20 WP	1.000 0.400 0.004	PE (3-5 DAT) POE (25 DAT) POE (21-25 DAT)	G S+BL S+BL
Ethoxysulfuron 15 WSG + Anilofos 30 EC	0.020 0.375	POE (10-12 DAT)	S+BL G
Clomazone 15 EC + Propanil 30 EC	0.150 0.300	POE (10 DAT)	G+S+BL
Clomazone 15 EC + Propanil 30 EC	0.200 0.400	POE (20 DAT)	G+S+BL
Oxadiazon 25 EC or Oxyfluorfen 23.5 EC + Finely powdered copper sulphate	0.750 0.150 10-15	PE (4-8 DAT) PE (4-8 DAT)	Algal weeds

G: Grass; S: Sedge; BL: Broad leaf; DAT: Days after transplanting; PE: Pre-emergence; POE: Post-emergence; fb. : Followed by; HW: Hand weeding

6.3. Wheat:

Different options for weed management in wheat are as follows:

a) Cultural: Use of higher seed rate (125-150 kg/ha) and closer row spacing (15 cm)

b) Mechanical: Hand weeding twice at 21 and 42 days after sowing.

c) Chemical: Herbicides found promising are presented in Table 11.

Table 11. Promising herbicides for weed management in wheat

Herbicide	Dose (Kg a.i./ha)	Time of application	Weed spectrum
Oxyfluorfen 23.5 EC + One HW	0.15 -	PE 21 DAS	G+S+BL
Sulfosulfuron 75 WDG	0.025	POE (30-35 DAS)	G+BL
Metsulfuron methyl 20 WP	0.004-0.006	POE (25-35 DAS)	BL
2,4-D Na Salt 80 WP	0.8	POE (30 DAS)	BL
Clodinofop propargyl 15 WP + 2,4-D Na salt 80 WP (tank mix)	0.060 0.500	POE (30-35 DAS)	G BL
Isoproturon 75 WP + 2,4-D Na salt 80 WP (tank mix)	0.75 0.6	POE (20-25 DAS)	G BL
Metoxuron 80 WP	1.0	POE (25-30 DAS)	G+BL
Metribuzin 70 WP	0.25-0.30	POE (35 DAS)	G+S+BL

G: Grass; S: Sedge; BL: Broad leaf; DAS: Days after sowing; PE: Pre-emergence; POE: Post-emergence; HW: Hand weeding

6.4. Jute:

Different options for jute weed management are as follows:

a) Cultural: sowing of pure seeds in line at right time.

b) Mechanical: About 40% of total cost of cultivation goes on manual weeding alone. Hand weeding twice at 21 and 35 days after sowing (DAS), and four wheel hoeings are recommended.

c) Integrated weed management (IWM): This is the best option for jute crop (Table 12) though Rajkhowa (2008) advocated the use of fenoxaprop-p-ethyl 9 EC as post-emergence (10 DAS) at 67.5 g/ha for safe control of the grassy weeds.

Table 12. Integrated weed management in jute

Chemical + Mechanical	Dose (Kg a.i./ha)	Time of application	Weed spectrum
Fluchloralin 45 EC + One HW	1.0 -	PPI (2 DBS) 21 DAS	G+BL S
Quizalofop ethyl 5 EC + One HW	0.04-0.05 -	POE (15-21 DAS) 35 DAS	G S+BL

G: Grass; S: Sedge; BL: Broad leaf; DAS: Days after sowing; DBS: Days before sowing; PPI: Pre-plant incorporation; POE: Post-emergence; HW: Hand weeding

6.5. Sugarcane:

Bhowmick (2004) suggested the following options for weed management in sugarcane:

a) Cultural: Deep summer ploughing and intercropping between sugarcane rows with cowpea, frenchbean, urdbean and soybean in spring planted crop, and gram and peas in autumn planted crop.

b) Mechanical: i). Two to three row cultivations besides two to three hand weedings at 30,60 and 90 days after planting (DAP).
ii). Mulching with 5-10 cm thick sugarcane trash at 45 DAP.

c) Chemical: Herbicides like 2, 4-D, triazines (simazine or atrazine) and metribuzin are generally recommended (Table 13).

Table 13. Promising herbicides for weed management in sugarcane

Herbicide	**Dose (Kg a.i./ha)**	**Time of application**	**Weed spectrum**
Simazine/Atrazine 50 WP	2.0-3.0	PE (2-3 DAP)	G+BL
Simazine 50 WP fb. 2,4-D Na salt 80 WP	2.0 2.0	PE (2-3 DAP) POE (60-70 DAP)	All including *C. rotundus*
2,4-D Na salt 80 WP fb. 2,4-D Na salt 80 WP + 3 hoeings starting from 25 DAP at 20 days interval	2.0 2.0 -	POE (10 DAP) POE (45-50 DAP) -	All
Metribuzin 70 WP	1.0-1.5	POE (21 DAP)	G+S+BL

G: Grass; S: Sedge; BL: Broad leaf; DAP: Days after planting; PE: Pre-emergence; POE: Post-emergence; fb: Followed by

6.6. Potato:

Different options for weed management are as follows:

a) Cultural: Thorough earthing up.

b) Mechanical: Hand weeding twice at 20 and 40 days after planting (DAP).

c) Chemical: Among different chemical herbicides, pre-emergence application of metribuzin 70 WP proved to be the most effective (Ghosh *et al.*, 1999 and 2002). However, the promising ones are presented in Table 14.

Table 14. Promising herbicides for weed management in potato

Herbicide	Dose (Kg a.i./ha)	Time of application	Weed spectrum
Linuron 50 WP	1.5	PE (5-6 DAP)	G+BL
Metribuzin 70 WP	0.5-0.8	PE (3 DAP)	G+S+BL
Pendimethalin 30 EC + Clomazone 15 EC (ready mix)	0.225-0.450	PE	G+S+BL

G: Grass; S: Sedge; BL: Broad leaf; DAP: Days after planting; PE: Pre-emergence

6.7. Maize:

Different options for weed management are as follows:

a) Cultural: Summer ploughing, timely sowing and proper spacing

b) Mechanical: Manual weeding at 2-3 and 6 weeks after sowing

c) Chemical: Herbicides found promising are presented in Table 15.

Table 15. Promising herbicides for weed management in maize

Herbicide	Dose (Kg a.i./ha)	Time of application	Weed spectrum
Simazine/Atrazine 50 WP	1.0-2.0	PE (1 DAS)	G+BL
Alachlor 50 EC	2.5	PE	G+BL

G: Grass; BL: Broad leaf; DAS: Days after sowing; PE: Pre-emergence

6.8. Pulses:

Bhowmick *et al.* (2009) reported that though hand weeding (HW) twice at 20 and 40 days after sowing (DAS) proved better in minimizing weed growth upto harvest and maximizing seed yield, it remained statistically at par with the pre-emergence (PE) application of pendimethalin and pre-plant incorporation (PPI) of fluchloralin at 1.0 kg/ha, when each supplemented with one HW at 40 DAS in greengram during pre-*kharif*. Similarly, integration of pendimethalin 30 EC at 0.75 kg/ha as PE with one HW at 40 DAs was found superior with regard to weed management and yield improvement in blackgram during *kharif* (Bhowmick and Gupta, 2005). In the same way, Bhowmick *et al.* (2008) recorded results of pendimethalin + imazethapyr (PPI) applied at 0.75 kg/ha in integration with one HW (30 DAS), being comparable with HW twice (30 and 60 DAS) in chickpea. Status of weed management in pulse crops has also earlier been reviewed by Saraswat and Mishra (1996). However, a list of promising herbicides for different pulse crops is given in Table 16.

Table 16. Promising herbicides for weed management in pulses

Crop	Mechanical / cultural	Dose (Kg a.i./ha)	Time of application	Weed spectrum
Soybean	Metribuzin 70 WP Fluchloralin 45 EC Trifluralin 48 EC Alachlor 50 EC Pendimethalin 30 EC Imazethapyr 10 WP Quizalofop ethyl 5 EC Fenoxaprop-p-ethyl Chlorimuron ethyl 25 WP HW twice	1.0 1.0 1.0 2.0 1.0 0.15 0.05 0.075 0.009 -	PE PPI PPI PE PE PE / POE POE POE POE 20 & 40 DAS	G+S+BL G+BL G+S+BL G+BL G+BL S+BL G G S+BL -
Green gram	Alachlor 50 EC + one HW Fluchloralin 45 EC + one HW Pendimethalin 30 EC + one HW HW twice	1.0 - 1.0 - 1.0 - -	PE 40 DAS PPI 40 DAS PE 40 DAS 20 & 40 DAS	G+BL G+BL G+BL G+S+BL
Black gram	Pendimethalin 30 EC+ One HW HW twice	0.75 - -	PE 40 DAS 20-25 & 40-45 DAS	G+BL S -
Pigeonpea	Alachlor 50 EC HW twice Intercropping with mung/kalai/cowpea	1.0 - -	PE 25 & 45 DAS -	G+BL - -
Cowpea	Alachlor 50 EC Oxyfluorfen 23.5 EC HW twice	1.0 0.09 -	PE PE 15 & 30 DAS	G+BL G+BL -
Lentil	Pendimethalin 30 EC Pendimethalin 30 EC+ HW HW twice	1.0 0.75 -	PE PE 40 DAS 30 & 60 DAS	G+BL G+BL G+S+BL

Table 16. (Continued)

Crop	Mechanical / cultural	Dose (Kg a.i./ha)	Time of application	Weed spectrum
Chickpea	Pendimethalin 30 EC +Imazethapyr 2 EC	1.0	PE	G+S+BL
	Pendimethalin 30 EC	1.5	PE	G+BL
	Fluchloralin 45 EC	1.5	PPI	G+BL
	Trifluralin 48 EC	1.0	PPI	G+S+BL
	HW twice	-	30 & 60 DAS	-
Chickpea	Pendimethalin 30 EC +Imazethapyr 2 EC	1.0	PE	G+S+BL
	Pendimethalin 30 EC	1.5	PE	G+BL
	Fluchloralin 45 EC	1.5	PPI	G+BL
	Trifluralin 48 EC	1.0	PPI	G+S+BL
	HW twice	-	30 & 60 DAS	-

G: Grass; S: Sedge; BL: Broad leaf; DAS: Days after sowing; PE: Pre-emergence; POE: Post-emergence; PPI: Pre-plant incorporation; HW: Hand weeding

6.9. Oilseeds:

Status of weed management in oilseed crops has earlier been reviewed by Saraswat and Mishra (1996).In groundnut, hand weeding twice at 20 and 40 days after sowing (DAS), and alachlor at 0.96 kg/ha as pre-emergence proved to be the best by registering maximum pod as well as oil yield because of decreased crop-weed competition, thereby providing congenial condition for better crop growth till harvest (Bhowmick *et al.*, 2001; Ghosh *et al.*, 2001). However, some promising herbicides for weed management in different oilseed crops are listed in Table 17.

Table 17. Promising herbicides for weed management in oilseeds

Crop	**Mechanical / cultural**	**Dose (Kg a.i./ha)**	**Time of application**	**Weed spectrum**
Groundnut	Alachlor 50 EC	1.0	PE (2 DAS)	G+BL
	Fluchloralin 45 EC +	1.5	PPI	G+BL
	HW	-	30-35 DAS	
	HW twice	-	20 & 40 DAS	G+S+BL
Rapeseed-Mustard	Nitrofen	1.0-2.0	PE	G+BL
	Isoproturon 75 WP	0.75	POE (25-30 DAS)	G
	Fluchloralin 45 EC	1.0-1.5	PPI	G+BL

Sesame	Alachlor 50 EC + One HW Fluchloralin 45 EC HW twice	1.5 - 1.0 -	PE 30 DAS PPI 15 & 30 DAS	G+BL S G+BL G+S+BL
Linseed	Isoproturon 75 WP MCPB / MCPA amine salt	0.75-1.0 0.5	PE/POE POE	G
Safflower	Alachlor 50 EC Fluchloralin 45 EC Oxadiazon 25 EC Trifluralin 48 EC HW twice	1.0-1.25 1.0 1.0 1.0 -	PE PPI PE PPI 25-30 & 45-50 DAP	G+BL G+BL G+BL G+S+BL G+S+BL
Sunflower	Nitrofen Alachlor 50 EC HW twice	2.0 2.0 -	PE PE 15 & 30 DAS	G+BL G+BL -
Castor	Alachlor 50 EC Fluchloralin 45 EC Trifluralin 48 EC	1.25 1.0 1.0	PE PPI PPI	G+BL G+BL G+S+BL
Niger	Pendimethalin 30 EC Fluchloralin 45 EC Removal of *Cuscuta* seeds by sieving before sowing or infested seedlings at early crop growth	1.0-1.25 1.0 -	PE PPI -	G+BL G+BL -

G: Grass; S: Sedge; BL: Broad leaf; DAS: Days after sowing; PE: Pre-emergence; POE: Post-emergence; PPI: Pre-plant incorporation; HW: Hand weeding

6.10. Vegetables:

Integrated weed management practices for vegetable crops have been briefed in Table 18.

Table 18. Integrated weed management (IWM) in vegetables

Crop	IWM	Dose (Kg a.i./ha)	Time of application	Weed spectrum
Cucurbits & transplanted tamato	Butachlor 50EC + One HW	2.0 -	PE (1-2 DAT) 5-6 WAS	G S+BL
Bean, Tomato, Okra	Alachlor 50EC + One HW Fluchloralin 45 EC + One HW	2.0 - 1.0 -	PE (1-2 DAT) 5-6 WAS PPI 5-6 WAS	G+BL - G+BL S

Table 18. (Continued)

Crop	IWM	Dose (Kg a.i./ha)	Time of application	Weed spectrum
Onion, garlic, brinjal, cole crops, transplanted tomato, pepper, carrot, green cropss	Fluchloralin 45 EC + One HW	1.0 -	PPI 5-6 WAS	G+BL S
Onion (transplanted & direct seeded)	Oxyfluorfen 23.5 EC	0.2-0.3	POE (10 DAT)	G+BL

G: Grass; BL: Broad leaf; DAT: Days after transplanting; WAS: Weeks after sowing; PE: Pre-emergence; POE: Post-emergence; PPI: Pre-plant incorporation; HW: Hand weeding

6.11. Non-crop areas:

Different options for weed management are as follows:

a) Mechanical: Though weeds in non-crop areas can be controlled by mowing, burning, grazing, discing and hand pulling, these are cost-prohibitive, labour-intensive and time-consuming (Bhowmick *et al.*, 2005).
b) Chemical : Glyphosate seems to be a safe and effective herbicide which can be used at low rate with improved formulation (Bhowmick *et al.*, 2005). Ammonium salt of glyphosate may have better performance and increased phytotoxicity in comparison to glyphosate (Duke, 1988; Ampong-Nyarko and De Datta, 1991). Glyphosate IPA salt + 2,4-D IPA salt (ready mix) shows excellent kill of diverse weed species within 20 days after spraying (Ghosh *et al.*, 2002). A list of herbicides recommended for such areas is given in Table 19.

Table 19. Promising herbicides for weed management in non-crop areas

Herbicide	**Dose**
Glyphosate 41 SL+ 2,4-D Na salt 80 WP (tank mix)	7.5 ml/l 5 g/l
Paraquat 25 WSC / Glyphosate 41 SL / Dalapon 50 WP + 2,4-D Na salt 80 WP	1.0 kg a.i./ha 1.0 kg a.i./ha
Glyphosate IPA salt 13.5% + 2,4-D IPA salt 13.5% (ready mix) (1.5%)	8 l/ha

7. ENVIRONMENTAL IMPACT, DEGRADATION AND TRANSFORMATION OF HERBICDES

Herbicides after application in soil, plant or aquatic system have got some impact either positive or negative in the environment. The positive impact is desirable for which purpose herbicide is applied for sustainable crop production. But negative impact of these chemicals and their degradative products are hazardous to human, animal as well as to environment. So, assessment of the harmful effect on the environment along with degradation, transformation and metabolism of the herbicides needs to be studied. Some of the degradation, transformation and metabolism studies in West Bengal as well as in India are discussed in the following paragraphs.

Pendimethalin [N-(1-ethylpropyl)-2, 6-dinitro-3, 4-xylidine,] (I) is a selective pre-emergent herbicide controlling most annual grasses and certain broad leaf weeds in cotton, paddy, soybean etc. Several soil fungi *Aspergillus flavus*, *A. terreus*, *Fusarium solani*, *F. oxysporoum*, *Penicillium citrinum* and *P. simlicissinum* effectively degraded pendimethalin in mineral solution. Degradation of pendimethalin (Barua *et al.,* 1990) by *F. solani* resulted in the characterization of three metabolites (Fig. 1) as N-propyl-3-methyl-4-hydroxy-2, 6-dinitroaniline (II), N-(1-ethylpropyl)-2-amino-6-nitro-3, 4-xylidine (III) and 2, 6-dinitro-3, 4-xylidene (IV).

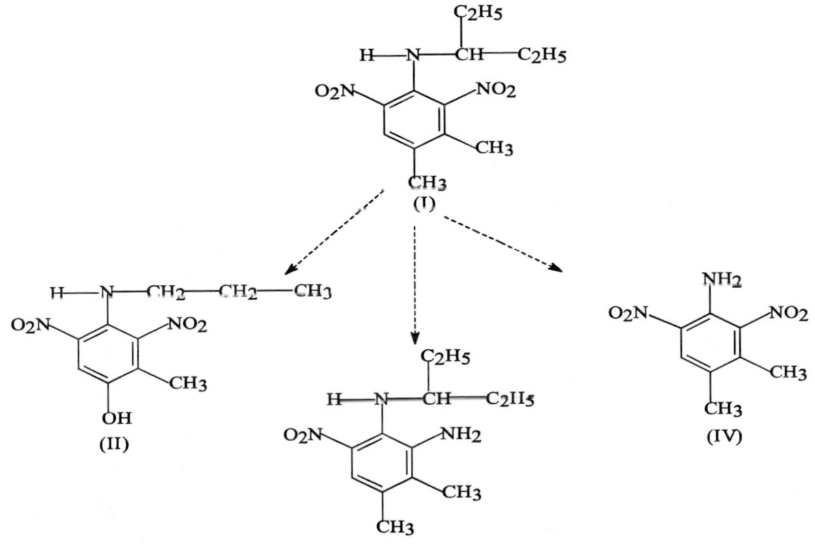

Fig.1. Hypothetical transformation pathway of pendimethalin by *F. solani*

Oxadiazon [2-tert-butyl-4-(2, 4-dichloro-5-isopropoxyphenyl)-Δ^2-1, 3, 4-oxadiazolin-5-one] is a selective pre-emergent soil applied herbicide, was possible to isolate *Fusarium solani* from soil cropped under paddy cultivation with no previous history and which was degraded by co-metabolic process and eleven metabolites were characterized based on IR, H^1NMR and GC-MS data. Only three of them were characterized (Chakraborty *et al.,* 1995) as M_1 [2,4-(dichloroisopropoxy)benzene], M_2 [1-(2,4-dichloro-5-isopropoxyphenyl)-1-

(methoxy-carbonyl)-1,2-trimethyl-acetyl hydrazine], and M_3 [1-trimethyl-acetyl-2-(2,4-dichloro-5-isopropoxyphenyl) hydrazine] as shown in Fig. 2.

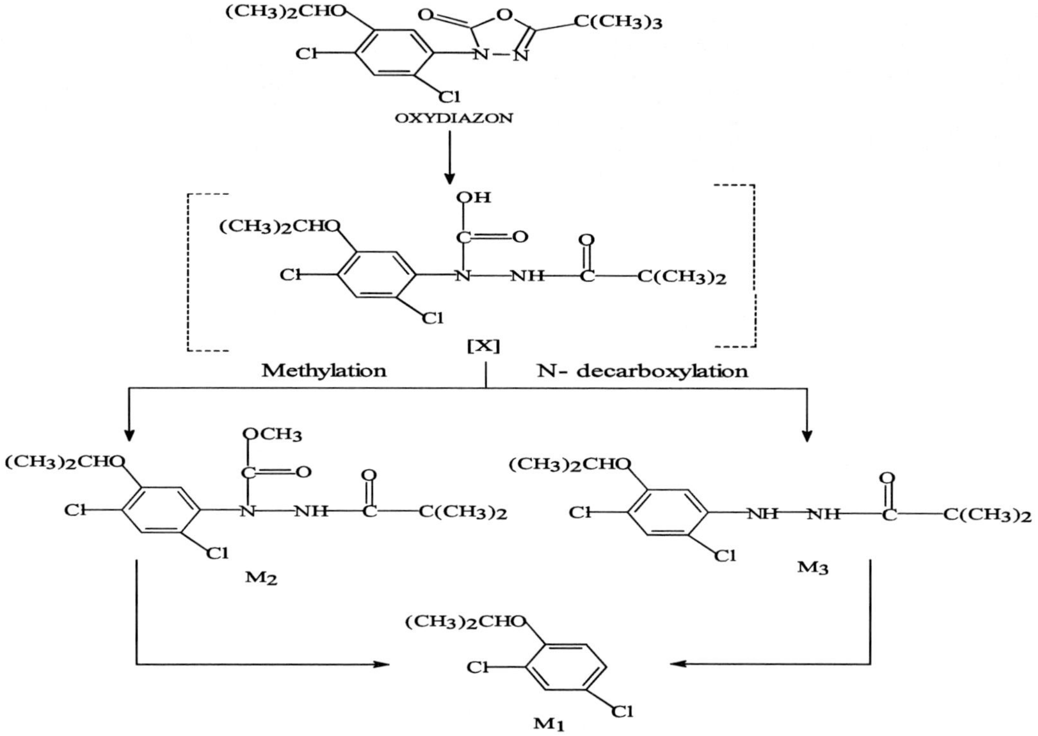

Fig.2. Hypothetical transformation pathway of oxadiazon by *F. solani.*

A laboratory study on oxyfluorfen [2-chloro-1-(3-ethoxy-4-nitrophenoxy)-4-(trifluoromethyl) benzene], a member of nitrodiphenyl ether herbicides, revealed that *A. chroococcum* degraded more than 60% of the herbicide in 7 days utilizing it as a sole cabon source, leading to the formation of couples of metabolites, of which two major metabolites (Fig. 3) were characterized, namely, metabolite I [N-[4-{2-chloro-4-(trifluoromethyl)-phenoxy}-2- ethoxyphenyl] acetamide}] and II [4-[2-chloro-4- (trifluoromethyl) phenoxy] -2-ethoxy benzene amine] during the microbial decomposition (Chakraborty *et al.,* 2002).

Studies on toxicokinetics, recovery and metabolism of triclopyr butolyl (ACTP) ester in goats showed that administrated ACTP ester was absorbed through the gastro intestinal tract of the goat. HPLC analysis revealed that all the tissues of different organs contained ACTP ester and its metabolites (Sar *et.al.* 2002). The proposed metabolic pathway of ACTP ester is given in Fig. 4.

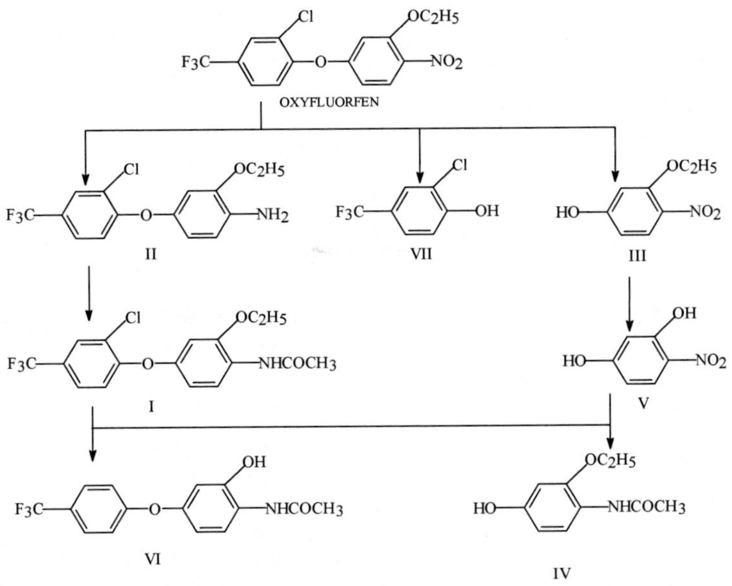

Fig.3. Hypothetical transformation pathway of oxyfluorefen by *A. chroococcum* (Beijerink)

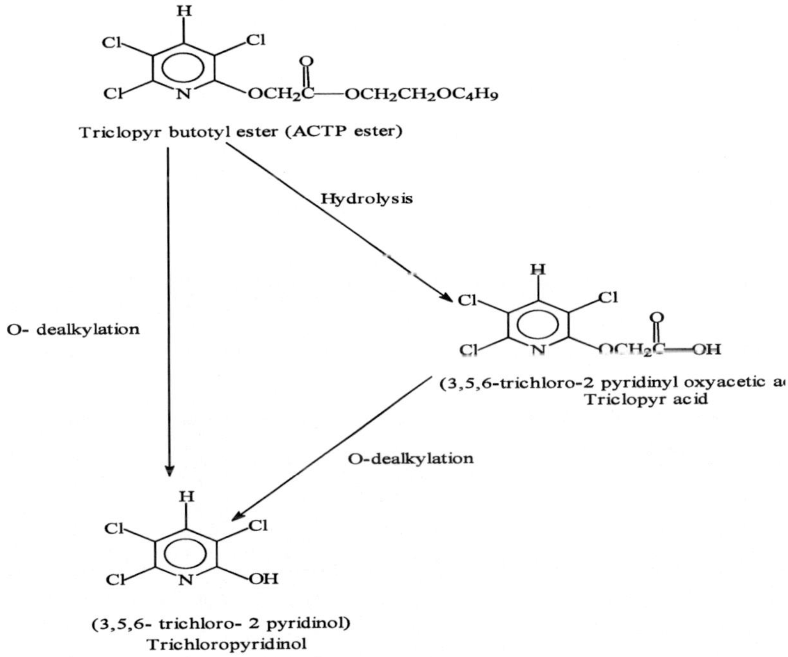

Fig.4. Probable metabolic pathway of triclopyr butolyl (ACTP) ester in goat

Toxico-kinetic behaviour and metabolism studies of metamitron and its effect on the cytochrome P_{450} content of liver microsomal pellets were carried out in black Bengal goat. Experiment revealed that repeated oral administration of metamitron at 30 mg kg^{-1} for 7 days caused induction of the cytochrome P_{450} content of liver microsomal pellets of goat, suggesting oxidative deamination (Fig. 5) of metamitron (Debnath *et al.*, 2003).

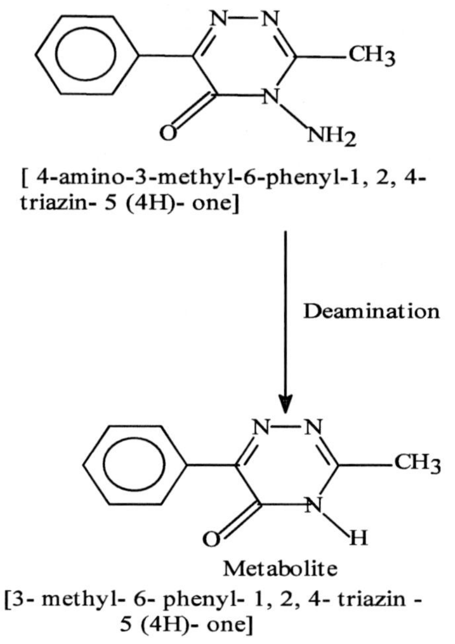

Fig.5. Probable metabolic pathway of metamitron

Dissipation of dichlobenil was studied in different soil types of West Bengal in different water regimes (Das and Chowdhury, 1998). The results revealed that the rate was higher in alluvial soil (about 92.90 to 95.50% lost after 60 days of application), while in laterite soil it was 4.60-9.30%. Dichlobenil might not pose any residual toxicity problem to the succeeding crop under the experimental conditions.

Laboratory simulated dissipation of metsulfuron methyl and chlorimuron ethyl in soils and their residual fate in rice, wheat and soybean at harvest was studied. The results revealed that the half-lives of metsulfuron methyl and chlorimuron ethyl ranged from 10.75 to 13.94 days, irrespective of soils and doses applied. Field trials with rice, wheat and soybean also revealed that these two herbicides could safely be recommended for application as no residues were detected in the harvest samples (Sanyal *et al.,* 2006).

The residue data of napropamide in field soil cropped with tea at different day's interval was studied. The corresponding dissipation rate, half-lives and regression equation were also calculated on the basis of residue data. It, in tea cropped soil, declined progressively with time as initial residues. The initial deposit of napropamide in tea cropped soil was found in the range of 1.18–1.49 and 2.08–2.90 μg/g for recommended dose and double the recommended dose, respectively, irrespective of any season and doses. The dissipation of napropamide in tea cropped soil followed the first-order kinetics with the half-life values varying from 12.54 to 27.87 days, irrespective of doses and season. In made tea, the initial concentration of napropamide was found in the range of 0.14–0.20 μg/g in recommended dose and 0.35–0.44 lg/g in double the recommended dose in three seasons. On 7th day after application the residue was found to be below detectable limit in both the treatments, irrespective of seasons. The results revealed that no residue of napropamide was detected in made tea at 7 days after application which would be befitting with the plucking schedule of the north-east region of India and it might not pose any residual toxicity in made tea (Biswas *et al.*, 2007).

8. RESEARCH PRIORITIES

In view of various issues and challenges confronting, there is a need for planning and orientation of research and development programmes for weed management to ensure sustainability in agricultural production. The major thrust areas in the research and development that deserve focus in future planning are as follows:

1. Weed biology (biological attributes that may be associated with survival and dispersal of species) and weed population dynamics (relating to weed seed germination, establishment and competition with crops);
2. Herbicide-moisture interaction to optimize herbicide use efficiency through factor adjusted dose recommendations;
3. Crop and herbicide rotation / sequence to address weed shift and weed resistance;
4. Herbicide mixture to enhance herbicide efficiency, lower optimum dose of herbicide, reduce cost of weed management, arrest weed shifts, prevent herbicide resistance in weeds and facilitate improvement in overall weed management;
5. Herbicide-insecticide/fungicide/fertilizer interaction for improving the efficacy and economy of weed management and minimizing the hazards to the crop and environment;
6. Herbicide safeners (use of chemicals that serve as antidotes to herbicides) to protect the crop against the toxic action of herbicides;
7. Herbicide residue (part of the herbicide left behind unused and undetoxified after use in soil) to understand the absorption, translocation and mode of action of herbicides as well as transformation and degradation of pathway in plants and soil;
8. Allelopathic effects of weeds and crops in relation to their competitiveness including ecological shifts in weed populations;
9. Weed management in the entire cropping system instead of the present concept of controlling weeds in a single crop;
10. Biological weed control including application of biotechnology for long-term suppression of weeds. Owing to have several advantages and disadvantages of bioherbicides, research work should be concentrated on their use for the control of weed species that escape standard chemical control. Besides, the important research priorities that need to be addressed in the near future in order to advance this field further (Charudattan, 1988; Yaduraju *et al.*, 2005) include: (i) development and registration of more bioherbicide candidates for economically important weeds, (ii) integration of bioherbicides with chemical pesticides and cultural practices, (iii) advancement in formulation, mass production, and delivery system for enhanced shelf-life and efficacy of bioagents, (iv) application of biotechnology for transformed bioherbicide agents for increased virulence and host-specificity, and (v) mixed formulation of microbial biopesticide agents and their integration with insects for enhanced broad-spectrum control of weeds;
11. Weed resistance to herbicides for continuous use of same or closely related herbicides in monoculture system, keeping an eye to prevent second generation problems as appeared leading to shifts in weed flora and to modify weed control programmes accordingly;
12. Crop management manipulation with a view to work out the influence of tillage, row spacing, crop canopy, seed rate, fertilizer and water management for all the crops;

13. Modeling of crop-weed interference (e.g. weed density-yield model as a basis for development of economic weed threshold model that can predict the weed density at which specific post-emergence weed control measures are warranted);

14. Herbicide application equipments to develop suitable sprayers, spray nozzles and granular application equipments besides enhancing their durability and reducing spraying cost;

15. Utilization of weeds for useful purposes and ethnobotany of weeds (e.g. possibility of using weeds like water hyacinth for compost making, biogas production, manufacture of paper pulp, board, mats, house roofs, fish food, food protein for non-ruminant animals, as a source of nitrogen and potassium for soil, and as mulching materials for weed control and conserving soil moisture, etc.); and

16. Transfer of weed management technology after working out the cost : benefit ratio based on a detailed socio-economic analysis.

9. Epilogue

Crop losses due to weeds are still very large and weed management cost is a major input cost for crop production in India. Herbicides including bioherbicides would be the number one agricultural inputs in very near future as because, the man power is going day by day costly and herbicides are economical to use in crop production with a larger profit margin. Besides, due to economic globalization, India would be the main food power in near future and of course, the productivity of all the crops including cereals, pulses and oilseeds should be enhanced. The marketing possibility of our agricultural and horticultural produces should be increased to the maximum extent possible through the use of modern inputs and improved mechanization. Here is the importance of herbicides in the field of weed management, especially under hot and humid climatic condition with good rainfall, usually prevailing in this subcontinent. But it seems appropriate to mention here that though chemical weeding effectively reduces the crop-weed competition and improves the farm labour efficiency, the herbicide chemicals being basically toxic in nature pose some concerns about food safety, environmental pollution, increased weed resistance to herbicides, destruction of beneficial organisms, etc. along with growing dependence on purchased input. Judicious use of herbicides, therefore, should be given the top priority. And lower rates of herbicides integrated with other methods of weed management are invariably more efficient, economical and sustainable than using higher dose of herbicides alone. In addition, it is desirable to initiate interdisciplinary thorough study on transformation, metabolism, and environmental as well as economic impact of herbicide chemicals in a particular agro-ecosystem before making out any recommendation on a sound chemical weed management practice. In other words, the very much need of the day is to undertake studies relating to weed biology (weed ecology, weed phenology, weed survey and weed taxonomy), weed physiology; weed microbiology, herbicide physiology, herbicide biochemistry and herbicide residue.

Due emphasis also needs to be given for utilizing frontier technologies towards different non-chemical methods of weed management starting from manoeuvring of crop cultivation systems to biological methods along with the efforts to develop bioherbicides. As low pathogen virulence and fastidious environmental conditions have been found to be the key restraints to bioherbicide development, efforts are also needed to overcome these in addition to the other limitations. Biological agents are most often host-specific and are usually limited to one or a few closely related weed species. As such, the narrow-spectrum bioherbicides cannot presently be considered as alternatives to broad-spectrum chemical herbicides and other weed control tactics, but should be considered as complementary adjuncts in integrated weed management (IWM) system. Thus, an IWM practice involving chemical herbicides as well as bioherbicides along with other tactics in judicious combination would be a cost-effective *vis-a-vis* eco-friendly remedy to the weed problems in Indian agriculture.

REFERENCES

Alagh, Y. K. 1988. Inaugural address on pesticides and Indian agriculture. *Natl. Sem. on "Changing pest situation in the current agriculture scenario of India"*, June 14-16, IARI, New Delhi.

Ampong-Nyarko, K. and De Datta, S. K. 1991. *A handbook of weed control in rice.* International Rice Research Institute, Los Banos, Philippines. p. 113.

Arora, S. and Gopal, M. 2004. Residue of pendimethalin after weed control in cabbage crop (*Brassica oleracea* L. var. *capitata*). *Bull. Environ. Contam. Toxicol.* 73: 106-110.

Auld, B. A., Hetherington, S. D. and Smith, H. E. 2003. Advances in bioherbicide formulation. *Weed Biol. Manag.* 3: 61-67.

Bansal, G. L. and Bhan, V. M. 1993. Status of research on allelopathy and future scope of work in India. *Indian Journal of Agricultural Sciences* 63 (12): 769-776.

Barua, A. S., Saha, J., Chaudhuri, S., Chowdhury, A. and Adityachaudhury, N. 1990. Degradation of pendimethalin by soil fungi. *Pestic. Sci.* 29 (4): 419-425.

Bhan, V. M. 1997. *Vision – 2000.* National Research Centre for Weed Science. Perspective Plan. 53 pp.

Bhan, V. M. and Singh, V. P. 1993. Integrated weed management (IWM) – An approach. (in) *Proc. Int. Symp.*, Nov. 18-20, Indian Society of Weed Science, Hisar. pp. 289-297.

Bhan, V. M. and Sushilkumar. 1998. Weed science research in India. *Indian J. Agril. Sciences* 68 (8, Spl. Issue): 567-582.

Bhowmick, M. K. 2000. *Parthenium hysterophorus* L. : A prolific weed of West Bengal and its management. *Science and Culture* 66 (7-8) : 277-278.

Bhowmick, M. K. 2001a. Impact of herbicides on energy utilization by high yielding summer paddy and associated weeds. *Annals of Plant Protection Sciences* 9 (1): 104-108.

Bhowmick, M. K. 2001b. Beware of the toxic weed *Argemone mexicana* L. *Science and Culture* 67 (5-6):179-180.

Bhowmick, M. K. 2002a. Optimization of pretilachlor dose for weed management in transplanted rice. *Annals of Plant Protection Sciences* 10 (1): 131-133.

Bhowmick, M. K. 2002b. Weed as alternate host of crop pest and its control measure. *Advances in Plant Sciences* 15 (1): 141-143.

Bhowmick, M. K. 2004. Weeds and their management in sugarcane. *SATSA Mukhapatra-Annual Technical Issue* 8: 47-55.

Bhowmick, M. K. and Biswas, A. 2007. Bioherbicide – A viable input for eco-friendly weed management. *SATSA Mukhapatra – Annual Technical Issue* 11: 29-43.

Bhowmick, M. K., Biswas, P. K. and Chowdhury, A. 2009. Integrated weed management in summer mungbean. *Abstract.* The 16th West Bengal State Science and Technology Congress, Feb. 28 - Mar. 01, University of Burdwan, West Bengal, India. p.545.

Bhowmick, M. K., Biswas, P. K., Saha, S. and Chowdhury, A. 2008. Integrated weed management in chickpea. *Abstract.* The 15th West Bengal State Science and Technology Congress, Feb. 28 - 29, 2008, Bengal Engineering and Science University, Shibpur, West Bengal, India. p.126.

Bhowmick, M. K., Ghosh, P., Ghosh, R. K. and Pal, A. K. 2005. Grain production and energy utilization due to pretilachlor application on transplanted *kharif* paddy. *Plant Archives* 5 (1): 139-143.

Bhowmick, M. K. and Ghosh, R. K. 2002a. Development and use of bioherbicides in the field of weed management. *Proc. Natl. Seminar on "Development and use of biofertilizers, biopesticides and organic manures,"* Nov. 10-12, 2000, B.C.K.V., Kalyani, West Bengal. pp. 237-245.

Bhowmick, M. K. and Ghosh, R. K. 2002b. Relative efficacy of herbicides against weed incidence in summer rice . *Advances in Plant Sciences* 15 (2): 499-503.

Bhowmick, M. K. and Ghosh, R.K. 2006. Sole and combined use of herbicides for weed control in transplanted rice during dry season . *Oryza* 43 (I): 545-47.

Bhowmick, M. K., Ghosh, R. K. and Ghosh, P. 2005. Management of weeds with new moleule XL 71 AG under non-crop situation. *Journal of crop and Weed* 1(1) : 70-72.

Bhowmick, M. K., Ghosh, R. K. and Pal, D. 2000. Bio-efficacy of new promising herbicides for weed management in summer rice. *Indian Journal of Weed Science* 32 (1&2): 35-38.

Bhowmick, M. K. and Gupta, S. 2005. Herbicidal-cum-integrated approach to weed management in urdbean . *Journal of Crop and Weed* 1 (2): 75-77.

Bhowmick, M. K., Ray, D. and Ghosh, R. K. 2001. Cultural and chemical weed management options for improving growth, nodulation and yield level in groundnut. *Proceedings. Natl. Sem. on "Frontiers of Crop Management"* (S. K. Mukhopadhyay, D. C. Ghosh and G. C. De, eds.), Feb. 1-3, 2001, Visva-Bharati, Sriniketan, West Bengal . pp.156-158.

Bhowmick, M. K. and Mandal, B. K. 2001. Biotechnological approaches to the management of weeds in Indian agriculture. *Science and Culture* 67 (9-10): 275-280.

Bhowmik, P. C. 1998. Herbicides in relation to food security and environment. A global perspective. *Proc. First Int. Agronomy Congress*, Nov. 23-27, New Delhi, India. pp.244-258.

Bhowmik, P. C. 1999. Herbicides in relation to food security and environment: A global perspective. *Indian J. Weed Sci.* 31 (3&4): 111-123.

Biswas, P. K., Pramanik, S. K., Mitra, S. R. and Bhattacharyya, A. 2007. Persistence of napropamide in/on tea under north-east Indian climatic condition. *Bull Environ Contam Toxicol* .79: 566–569.

Bowers, R. C. 1986. Commercialization of collego - An industrialist's view. *Weed Sci.* 34 (Suppl.1): 24-25.

Boyette, C. D. and Abbas, H. K. 1995. Weed control with mycoherbicides and phytotoxins. *ACS Symp. Series: Allelopathy Organisms–Process and Application.* American Chemical Society. pp. 280-299.

Burnett, H. C., Tucker, D. P. H. and Patterson, M. E. 1973. Biological control of milkweed vine with a race of *Phytophthora citrophthora. Proc. Florida Horticultural Soc.* 86: 111-115.

Buttar, G. S. and Aggarwal, N. 2005. Bioherbicides – A review. *Pestology* XXIX (6): 27-35.

Chakraborty, S. K., Bhattacharyya, A. and Chowdhury, A. 2002. Degradation of oxyfluorfen by *Azotobacter chroococcum* (Beijerink). *Bull. Environ. Contam. Toxicol.* 69: 203-209.

Chakraborty, S. K., Chowdhury, A., Bhattacharyya, A., Ghosh, S., Pan, S., Waters, R. and Adityachaudhury, N. 1995. Microbial degradation of oxadiazon by soil fungus *Fusarium. J. Agric. Food Chem.* 43: 2964-2969.

Charudattan, R. 1984. Microbial control of plant pathogens and weeds. *J. Georgia Entomol. Soc.* 19(3): 40-62.

Charudattan, R. 1988. Assessment of efficacy of mycoherbicide candidates. *Proc. The 7th International Symp. on "Biological control of weeds"*, Rome, Mar. 6-11.

Charudattan, R. 2001. Biological control of weeds by means of plant pathogens: Significance for integrated weed management in modern agro-ecology. *Biocontrol* 46: 229-260.

Charudattan, R. and Dinoor, A. 2000. Biological control of weeds using plant pathogens: Accomplishments and limitations. *Crop Bot.* 19: 691-695.

Charudattan, R., Linda, S. B., Kluepfel, M. and Osman,Y.A. 1985. Biocontrol efficacy of *Cercospora rodmanii* on water hyacinth. *Phytopathology* 75: 1263-1269.

Charudattan, R. and Walker, H. L. 1982. *Biological control of weeds with plant pathogens.* Wiley-Interscience, New York.

Chu, E. Y. I. 1983. Use of tissue culture for breeding herbicide tolerant varieties. *Proc. Workshop "Cell and tissue culture techniques for cereal crop improvement", IRRI,* Philippines. pp.303-315.

Das, A. K. and Chowdhury, A. 1998. Dissipation of dichlobenil in different soil types of West Bengal. *Pestology* 22 (6): 50-53.

Debnath, S. C., Kundu, A., Das, S. K., Mandal, T. K., Bhattacharyya, A., Chowdhury, A. and Chakraborty, A. K. 2003. Toxicokinetics, recovery and metabolism of metamitron in goat. *J. Agric. Food Chem.* 51 (20): 5977-5981.

Doraiswamy, Sabitha and Manickam, K. 1996. Biological control of weeds using plant pathogens - Prospects and future. *Advances in weed management in an agro-ecological context* (Summer Inst. -Short Course, June 10-19), TNAU, Coimbatore. pp.115-120.

Duke, S. O. 1986. Naturally occurring chemical compounds as herbicides. *Rev.Weed Sci.* 2: 17-44.

Duke, S. O. 1988. Glyphosate. (in) *Herbicidal chemistry, degradation and mode of action* 3 (P. C. Kearney and D. D. Kaufman, eds.). Mercel Dekker Inc., New York and Basel. pp. 1-70.

Gautam, K. C. and Mishra, J. S. 1995. Problems, prospects and new approaches in weed management. *Pesticides Information* XXI (1): 7-19.

Ghorbani, R., Leifert, C. and Seel, W. 2005. Biological control of weeds with antagonistic plant pathogens. *Advances in Agronomy* 86: 191-225.

Ghosh, P., Ghosh, R. K. and Bhowmick, M. K. 2002. Bio-efficacy of mix formulation of clomazone and propanil for weed management in transplanted *kharif* paddy. *Environment and Ecology* 20 (1): 147-151.

Ghosh , P., Mukherjee , N., Bhowmick, M. K. and Ghosh, R. K. 2005. Eco-friendly chemical weed management in transplanted *kharif* rice. (in) *Extended summaries. Natl. Biennial Conf.*, April 6-9, Indian Society of Weed Science, Department of Agronomy and Agrometeorology, PAU, Ludhiana: 15-16.

Ghosh, R. K. and Bhowmick, M. K. 2008. Strategies for ecorestoration of soil and water resources: Towards efficient input use *vis-à-vis* sustainable crop production. Invited paper. *SATSA Mukhapatra – Annual Technical Issue* 12: 23-29.

Ghosh, R. K., Bhowmick, M. K., Das, S. C. and Jana, P. K. 2001. Herbicidal management of weeds in groundnut (*Arachis hypogaea L.*). *Journal of Oilseeds Research* 18 (2): 195-197.

Ghosh, R. K., Bhowmick, M. K. and Ghosh, P. 2002. Efficacy of glyphosate and 2,4-D combinations for weed management in non crop areas. *Environment and Ecology* 20 (1): 143-146.

Ghosh, R. K., Bhowmick, M. K., Ghosh, P. and Saha, S. 2002. Bio-efficacy of metribuzin 70WP on weed control in potato (*Solanum tuberosum* L). *Abstracts of invited and contributed papers. Int. Conf. on "Vegetables for sustainable food and nutritional security in the new millennium"*, Nov. 11-14 (Org. by PNASF, Bangalore, ISVS, New Delhi and IIHR, Bangalore), Bangalore. pp. 257-258

Ghosh, R. K., Mitra, S, and Bhowmick, M. K. 1999. Weed status of potato in alluvial belts of West Bengal and its suitable management. *Indian Journal of Weed Science* 31 (1&2): 84-88.

Gohbara, M. and Yamaguchi, K. 1993. Biological agents for the control of paddy weeds in Japan. *Extension Bulletin No.369*. Food and Fertilizer Technology Center, Taipai City, Republic of China on Taiwan.

Greaves, M. P., Holloway, P. J. and Auld, B. A. 1999. Formulation of microbial herbicides. (in) *Formulation of microbial herbicides: Beneficial organisms and nematodes* (H. D. Burges, ed.). Kluwer Academic, Dordrecht, The Netherlands. pp. 203-233.

Gupta, V. P., Kumar, V., Mishra, R. K., Thiagarajan, V. and Datta, R. K. 2002. *Puccinia romagnoliana* Marie & Sacc.– a potential bioherbicide agent for biocontrol of purple nutsedge (*Cyperus rotundus* L.) in mulberry. *J. Phytopathol.* 150: 263-270.

Hatzios, K. K. 1987. Biotechnology applications in weed management: Now and in the future. *Advances in Agronomy* 41: 325-375.

Hoagland, R. E. 1990. Microbes and microbial products as herbicides.*ACS Symposium Series,* U.S.A.

Hodgson, R. H., Wymore, L. A., Watson, A. K., Snyder, R. H. and Collette, A. 1988. Efficacy of *Colletotrichum coccodes* and thidiazuron for velvetleaf (*Abutilon theophrasti*) control in soybean (*Glycine max). Weed Technol.*2: 473-480.

Islam, T., Bhowmick, M. K.,Ghosh, R. K. and Sounda, G. 2001. Effect of pretilachlor on weed control and yield of transplanted paddy. *Environment* and *Ecology* 19 (2): 265-268.

Johnson, B. J. 1994. Biological control of annual bluegrass with *Xanthomonas campestris* pv. *poannua* in bermudagrass. *Hortscience* 29: 659-662.

Karuppiah, R. and Seetharaman, K. 2004. Approach to sedge weeds management in rice ecosystem with rust and smut as bioherbicides. *J. Mycol. Pl. Pathol.* 34 (3): 948-952.

Kenney, D. S. 1986. DeVine –The way it was developed – An industrialist's view. *Weed Sci.* 34 (Suppl.1): 15-16.

Khachatourians, G. G. 1986. *Trends Biotechnol.* 4: 120-124.

Kocher, H. 1983. Influence of environmental factors on herbicide performance and crop and yield biology. (in) *Aspects of Applied Biology* 4. Association of Applied Biologists, Oxford. pp. 227-234.

Kremer, R. J. 2002. Bioherbicides: Potential successful strategies for weed control. (in) *Microbial biopesticides* (Koul, O. and Dhaliwal, G. S., eds.), Taylor and Francis Pub., London.

Mase, S. 1984. *Japan Pestic. Inf.* 45 : 27-30.

Misato, T. and Yamaguchi, I. 1984. *Outlook Agric.* 13 : 136-139.

Mukhopadhyay, S. K. 1986.Aquatic weed control in India. Paper presented in Indo-Pak-US Weed Control Workshop, Jaipur. *Pesticides*. Nov. ,1986. pp. 15-22.

Mukhopadhyay, S. K. 1992. Emerging problems and advances in weed management. *Proc.79th Indian Science Congress. Part II. Presidential Address* (Agric. Section), Baroda. pp. 1-16.

Mukhopadhyay, S. K. 1993. Scope and limitation of the application of biotechnology in integrated weed management. *Proc. Int. Symp. Indian Society of Weed Science*, Hisar, Nov.18-20, Vol.I: 221-224.

Mukhopadhyay, S. K. 1997. Weed management in changing agricultural situation–in retrospect and prospect. (in) *Proc. Symp. on "Pest management in changing agricultural situation"* (Dasgupta, M. K., ed.), Mar. 15-16, Palli Siksha Bhavan, Visva-Bharati. pp. 60-68.

Munakata, K., Yamada, O., Ishida, S., Futatsuya, F. and Yamamoto, H. 1973. (in) *Proc. Asian-Pacif. Weed Sci. Soc. Conf.* 4 : 215-219.

Oerke, E. C., Dehne, H. W., Schonbeck, F. and Weber, A. 1994. *Crop production and crop protection: Estimated losses in major food and cash crop*. Elsevier. 808 pp.

Omura, S., Hinotozawa, K., Imamura, N., and Murata, M. 1984. *J. Antibiot.* 37 : 939-940.

Pal, D., Bhowmick, M. K. and Ghosh, R. K. 2002 . New molecules for weed management in transplanted *kharif* paddy. *Environment and Ecology* 20 (1): 136-138.

Parker, C. and Fryer, J. D. 1975. Weed control problems causing major reduction in world food supplies. *FAO Plant Protection Bull.* 23 : 83-95.

Quimby, P. C., King, L. R. and Grey, W. E. 2002. Biological control as a means of enhancing the sustainability of crop/land management systems. *Agricult. Ecosyst. Environ.* 88:147-152.

Rajkhowa, D. J. 2008. Bioefficacy of fenoxaprop-p-ethyl (Whip Super 9 EC) in jute. *Pestology* XXXII (8): 44-46.

Ramadhar. 2008. Current crisis due to confluence of factors. *The Hindu Survey of Indian Agriculture*, National Press, Chennai, India: 14-17.

Ridings, W. H. 1986. Biological control of strangler vine in citrus–A researcher's view. *Weed Sci.* 34 (Suppl.1): 31-32.

Ridings, W. H., Mitchell, D. J., Schoulties, C. L. and El-Gholl, N. E. 1976. Biological control of milkweed vine in Florida citrus groves with a pathotype of *Phytophthora palmivora*. *Proc. IV Int. Symp. Biol. Control Weeds*, Gainesville, FL. pp. 224-240.

Ridings, W. H., Schoulties, C. L., El-Gholl, N. E. and Mitchell, D. J. 1977. The milkweed vine pathotype of *Phytophthora citrophthora* as a biological control agent of *Morrenia odorata*. *Proc.Int. Soc. Citriculture* 3: 877-881.

Rosskoph, E. N., Charudattan, R. and Kadir, J. B. 1999. Use of plant pathogens in weed control. (in) *Handbook of biological control* (Ed., Bellows and Fischer). Academic Press.

Sanyal, N., Pramanik, S. K., Pal, R. and Chowdhury, A. 2006. Laboratory simulated dissipation of metsulfuron methyl and chlorimuron ethyl in soils and their residual fate in rice, wheat and soybean at harvest. *Journal of Zhejiang University Science B* 7 (3): 202-208.

Sar, T. K., Bagchi, B., Das, S. K., Mandal, T. K., Chakraborty, A. K., Bhattacharyya, A. and Chowdhury, A. 2002. Toxicokinetics, recovery and metabolism of triclopyr butolyl (ACTP) ester in goats. *J. Agric. Food Chem.* 50 (15): 4202-4209.

Saraswat, V. N. and Mishra, J. S. 1996. Status of weed management research in pulse crops. *Pesticides Information* (Apr.-June): 11-17.

Saraswat, V. N. and Mishra, J. S. 1996. Weed management research in oilseed crops in India - An overview. *Pesticides Information* (July-Sept.): 31-37.

Saraswat, V. N. and Mitra, P. C. 1996. Weed management in jute and allied fibres. *Indian Farming* XXVI (12): 50-52.

Scheepens, P. C., Muller-Scharer, S. and Kempenaar, C. 2001. Opporôtnities for biological weed c_ntrol in Europe. *Biocontrol* 46: 127-138.

Scheepens, P. C. and van Zon, H. C. J. 1982. (in) *Microbial and viral pesticides* (E. Kuratak, ed.).Dekker, New York. pp. 623-641.

Scher, F. M. and Castagno, J. R. 1986. Biocontrol: A viewpoint from industry. *Can. J Plant Pathol.* 8: 222-224.

Singh, A. 2005. Ensuring food security (in) *Integrated Pest Management–Principles and applications*.Vol. 1 (Singh, A., Sharma, O.P., and Garg, D.K., eds.), CBS Publishers and Disô»4±:°ors, New Delhi: 1-8.

Singh, Sahadeva. 2001. Integrated weed management. *Employment News* XXV (53): 1-2.

Singh, S. P. 2001. Use of biological control methods in weed management from the point of eco-safety.(in) *Lead papers and abstracts. First Biennial Conf. in the new millennium on "Eco-friendly weed management options for sustainable agriculture*". Org. by ISWS & UAS (Bangalore), May 23-24, Bangalore: 23-35.

Smith, R. J. Jr. 1986. Biological control of northern jointvetch (*Aeschynomene virginica*) in rice and sopybean–A researcher's view. *Weed Sci.* 34 (Suppl.1): 17-23.

Tachibana, K. 1987. (in*) Pesticide Science and Biotechnology, 6th IUPAC Congress of Pesticide Chemistry* (R. Greenhalgh and T.R. Roberts, eds.). Blackwell, Oxford. pp. 145-148.

TeBeest, D. O. 1991. *Microbial control of weeds*. Chapman and Hall, New York.

TeBeest, D. O. 1996. Biological control of weeds with plant pathogens and microbial pesticides. *Advances in Agronomy* 56: 115-137.

Templeton, G.E. and Smith, R. J. Jr. 1977. (in) *Plant Disease: An Advanced Treatise* Vol. 1 (Hosfall, J. G. and Cowling, E. B., eds.). Academic Press, New York. pp. 167-176.

Templeton, G. E., Smith, R. J. Jr. and TeBeest, D. O.1986. *Rev. Weed Sci.* 2: 1-14.

Templeton, G. E., TeBeest, D. O. and Smith, R. J. Jr. 1979. Biological weed control with mycoherbicides. *Ann. Rev. Phytopathol.* 17: 301-310.

TeBeest, D. O. and Templeton, G. E. 1985. Mycoherbicides: Progress in the biological control of weeds. *Plant Disease* 69: 6-10.

Upadhyay, U. C. 1985. Weed management in oilseed crops. (in) *Oilseed production: Constraints and opportunities* (Srivastava, H. C., Bhaskaran, S., Vatsya, B. and Menon, K. K. G.). Oxford and IBH Pub. Co., New Delhi: 491-499.

Watson, A. K. 1989. Current advances in bioherbicide research. *http://www.eap.mcgill.ca /PCBCW_3 .html.*

Wymore, L. A., Watson, A. K. and Gotleib, A. R. 1987. Interaction between *Colletotrichum coccodes* and thidiazuron for control of velvetleaf (*Abutilon theophrasti*). *Weed Sci.* 35: 377-383.

Yaduraju, N. T. 2002. Weed management in oilseed crops. (in) *Oilseeds and oils: Research and development needs* (Rai, M., Singh, H., and Hegde, D. M., eds.). Indian Society of Oilseeds Research, Rajendranagar, Hyderabad: 172-183.

Yaduraju, N. T., Bhanu, C. and Sushilkumar. 2005. Biological control of weeds: Potential and prospects. (in) *Abstract Book. First Int. Weed Sci. Sem. on "Innovative approaches for eco-safety weed management"*, Org. by WBWSS, BCKV and EIPA, Jan.21-24. Kolkata, India: I-XII.

Yaduraju, N. T. and Mishra, J. S. 2005. Status and prospects of integrated pest management strategies for obnoxious weeds. (in) *Integrated Pest Management–Principles and applications*. Vol.1 (Singh, A., Sharma,O. P., and Garg, D. K., eds.), CBS Publishers and Distributors, New Delhi: 319-360.

INDEX

A

B

C

D

E

F

G

H

I

J

K

L

M

N

O

P

Q

R

S

T

U

V

W

Y

Z